Lecture Notes in Computer Science 11890

More information about this series at http://www.springer.com/series/8851

Ngoc Thanh Nguyen · Ryszard Kowalczyk ·
Jacek Mercik · Anna Motylska-Kuźma (Eds.)

Transactions on Computational Collective Intelligence XXXIV

Springer

Editor-in-Chief
Ngoc Thanh Nguyen
Wroclaw University of Technology
Wroclaw, Poland

Co-Editor-in-Chief
Ryszard Kowalczyk
Swinburne University of Technology
Hawthorn, VIC, Australia

Guest Editors
Jacek Mercik
WSB University in Wroclaw
Wroclaw, Poland

Anna Motylska-Kuźma
WSB University in Wroclaw
Wroclaw, Poland

ISSN 0302-9743   ISSN 1611-3349 (electronic)
Lecture Notes in Computer Science
ISSN 2190-9288   ISSN 2511-6053 (electronic)
Transactions on Computational Collective Intelligence
ISBN 978-3-662-60554-7   ISBN 978-3-662-60555-4 (eBook)
https://doi.org/10.1007/978-3-662-60555-4

This Springer imprint is published by the registered company Springer-Verlag GmbH, DE
part of Springer Nature
The registered company address is: Heidelberger Platz 3, 14197 Berlin, Germany

Transactions on Computational Collective Intelligence XXXIV

Preface

It is our pleasure to present to you the XXXIV volume of LNCS *Transactions on Computational Collective Intelligence* (TCCI). In Autumn 2018 (November 23) at the WSB University in Wroclaw, Poland, there was the fourth seminar on "Quantitative Methods of Group Decision Making." Thanks to the WSB University in Wroclaw we had an excellent opportunity to organize and financially support the seminar. This volume presents post-seminar papers of participants to this seminar. During the seminar we listened to and discussed over 18 presentations from 17 universities. The 34th issue of TCCI contains 12 high-quality, carefully reviewed papers.

The first paper "A Probabilistic Model for Detecting Gerrymandering in Partially-Contested Multiparty Elections" by Dariusz Stolicki, Wojciech Słomczyński, and Jarosław Flis is devoted[1] to finding a method for detecting gerrymandering in multiparty partially-contested elections, such as the Polish local election of 2014. A new method for detecting electoral bias, based on the assumption that voting is a stochastic process described by Polya's urn model, is devised. Since the partially-contested character of the election makes it difficult to estimate parameters of the urn model, an ad-hoc procedure for estimating those parameters in a manner untainted by potential gerrymandering is proposed.

In the second paper entitled "Power in Networks: A PGI Analysis of Krackhardt's Kite Network" by Manfred J. Holler and Florian Rupp one may find application of power index analysis to the well-known Krackhardt's kite social network by imposing a weighted voting game on the given network structure. It compares the results of this analysis, derived by applying the Public Good Index and the Public Value, with the outcome of employing the centrality concepts – degree centrality, closeness centrality, and betweenness centrality – that they find in Krackhardt (1990), and eigenvector centrality. The conclusion is that traditional centrality measures are rather a first approximation for evaluating the power in a network as they are considerably abstract from decision making and thereby of possible coalitions and actions. Power index analysis takes care of decision making, however, in the rather abstract (a priori) form potentially forming coalitions.

In the third paper "Orders of Criticality in Graph Connection Games" by Marco Dall'Aglio, Vito Fragnelli, and Stefano Moretti one may find the analysis of order of criticality of a player in a simple game and two indices inspired by the reasoning *a la* Shapley and *a la* Banzhaf mainly having in mind voting situations. Here, they devote our attention to graph connection games, and to the computation of the order of criticality of a player. The indices introduced may be used as centrality measures of the edges in preserving the connection of a graph.

---

[1] Hereafter description of the papers are directly taken from summaries prepared by their authors.

The fourth paper "The Capacity of Companies to Create an Early Warning System for Unexpected Events – An Explorative Study" by Johannes Platje presents a discussion of determinants of the capacity of companies to deal with unexpected events and an approach to the creation of a company's Early Warning System. Capacity determinants discussed include: lack of functional stupidity, paradigms, general trust, and awareness of fragility indicators. The results of research based on an explorative questionnaire are presented for two small Swiss and German companies. The working hypothesis for the research is that flatter organizational structures possess higher capacity to create an Early Warning System than more hierarchical organizational structures. There is some weak evidence confirming this hypothesis.

In the fifth paper entitled "Electoral Reform and Social Choice Theory: Piecemeal Engineering and Selective Memory" Hannu Nurmi observes that most electoral reforms are dictated by recognized problems discovered in the existing procedures or – perhaps more often – by an attempt to consolidate power distributions. Very rarely, if ever, is the motivation derived from the social choice theory even though it deals with issues pertaining to what is possible and what is impossible to achieve by using given procedures in general. He discusses some reforms focusing particularly on a relatively recent one proposed by Eric Maskin and Amartya Sen. It differs from many of its predecessors in invoking social choice considerations in proposing a new system of electing representatives. At the same time it exemplies the tradeoffs involved in abandoning existing systems and adopting new ones.

In the sixth paper "Repeated Trust Game – Statistical Results Concerning Time of Reaction" Anna Motylska-Kuźma, Jacek Mercik, and Aleksander Buczek present basic results regarding probability distributions together with the parameters related to the decision-making time in the repeated trust game. The results obtained are of a general nature, related to the waiting time for a reaction in computer-aided systems, as well as a special one related to the characteristics of the decision-makers participating in the experiment.

In the seventh paper entitled "Labeled Network Allocation Problems. An Application to Transport Systems" Encarnación Algaba, Vito Fragnelli, Natividad Llorca, and Joaquín Sánchez-Soriano investigate networks in which there are more than one arc connecting two nodes. These multiple arcs connecting two nodes are labeled in order to differentiate each other. Likewise, there is traffic or flow among the nodes of the network. The links can have different meanings as such roads, wire connections, or social relationships; and the traffic can be for example passengers, information, or commodities. When we consider that labels of a network are controlled or owned by different agents then we can analyze how the worth (cost, profit, revenues, power, etc.) associated with the network can be allocated to the agents. The Shapley quota allocation mechanism is proposed and characterized by using reasonable properties. Finally, in order to illustrate the advantages of this approach and the Shapley quota allocation mechanism, an application to the case of the Metropolitan Consortium of Seville is outlined.

The eighth paper entitled "Seat Apportionment by Population and Contribution in European Parliament After Brexit" by Cesarino Bertini, Gianfranco Gambarelli, Izabella Stach, and Giuliana Zibetti present the problem of apportioning seats to member countries of the European Parliament after Brexit and in view of new accessions/exits, as countries with strong economies (and their consequent large contributions to the European Union) require that they have greater representative weight in the European Parliament. In this paper, they propose a model for seat apportionment in the European Parliament, which assigns seats taking into account both the percentages of the populations and the percentages of the contributions by each member state to the European Union budget by means of a linear combination of these two quantities. The proposed model is a modification of the approach given by Bertini, Gambarelli, and Stach in 2005. Using the new model, they studied the power position of each European Union member state before and after the exit of the United Kingdom using the Banzhaf power index.

In the ninth paper "The Use of Group Decision-Making to Improve the Monitoring of Air Quality" of Cezary Orłowski, Piotr Cofta, Mariusz Wąsik, Piotr Welfler, and Józef Pastuszka present the use of methods supporting group decision making for the construction of air quality measurement networks. Their article presents a case study of making group decisions related to the construction of a hybrid network for measuring air quality in city of Gdańsk. Two different methods of data processing were used in the decision making process. The first one is using fuzzy modeling for quantitative data processing to assess the quality of PM10 measurement data. The other is using trust metrics for the IoT nodes of four different measurement networks. The presented example shows the complexity of the decision-making process itself as well as the choice of the method. The authors deliberately used both the quantitative and qualitative methods in the decision-making process to show the need to search for the right method by decision-makers.

The tenth paper entitled "Bi-proportional Apportionments" is written by Mirko Bezzi, Gianfranco Gambarelli, and Giuliana Zibetti. In the article an apportionment method is proposed that generalizes Hamilton's method for matrices, optimizing proportionality in both directions, both for rows and columns. The resulting matrix respects fixed totals for rows and columns even when such totals do not satisfy standard criteria (monotonicity, maximum, or minimum of Hare), for example following the allocation of majority prizes to parties or coalitions. Optionally, if required, the result can also respect the minimum Hare quotae for rows and columns.

The eleventh paper is the joint work of Josep Freixas and Montserrat Pons. The paper is entitled "A Probabilistic Unified Approach for Power Indices in Simple Games." Many power indices on simple games have been defined trying to measure, under different points of view, the *a priori* "importance" of a voter in a collective binary voting scenario. A unified probabilistic way to define some of these power indices is considered in this paper and it is also shown that six well-known power indices are obtained under such a probabilistic approach. Moreover, some new power indices can naturally be obtained in this way.

The twelfth paper "The Story of the *Poor* Public Good Index" is written by Manfred Holler. His paper starts from the hypothesis that the public good index (PGI) could be much more successful if it were introduced by a more prominent game theorist. He argues that the violation of local monotonicity, inherent to this measure of a priori voting power, can be an asset – especially if the public good interpretation is taken into consideration and the PGI is (re-)assigned to I-power, instead of P-power.

We would like to thank all authors for their valuable contributions to this issue and all reviewers for their opinions which helped to keep the papers in high quality. Our very special thanks go to Prof. Ngoc-Thanh Nguyen who encouraged us to prepare this volume and who helps us to publish this issue in due time and in good order.

August 2019   Jacek Mercik
                                                Anna Motylska-Kuźma

Transactions on Computational Collective Intelligence

This Springer journal focuses on research in applications of the computer-based methods of computational collective intelligence (CCI) and their applications in a wide range of fields such as the Semantic Web, social networks, and multi-agent systems. It aims to provide a forum for the presentation of scientific research and technological achievements accomplished by the international community.

The topics addressed by this journal include all solutions of real-life problems for which it is necessary to use CCI technologies to achieve effective results. The emphasis of the papers published is on novel and original research and technological advancements. Special features on specific topics are welcome.

Contents

A Probabilistic Model for Detecting Gerrymandering
in Partially-Contested Multiparty Elections 1
   Dariusz Stolicki, Wojciech Słomczyński, and Jarosław Flis

Power in Networks: A PGI Analysis of Krackhardt's Kite Network 21
   Manfred J. Holler and Florian Rupp

Orders of Criticality in Graph Connection Games 35
   Marco Dall'Aglio, Vito Fragnelli, and Stefano Moretti

The Capacity of Companies to Create an Early Warning System
for Unexpected Events – An Explorative Study 47
   Johannes (Joost) Platje

Electoral Reform and Social Choice Theory: Piecemeal Engineering
and Selective Memory 63
   Hannu Nurmi

Repeated Trust Game – Statistical Results Concerning Time of Reaction 74
   Anna Motylska-Kuźma, Jacek Mercik, and Aleksander Buczek

Labeled Network Allocation Problems. An Application
to Transport Systems 90
   *Encarnación Algaba, Vito Fragnelli, Natividad Llorca,
and Joaquin Sánchez-Soriano*

Seat Apportionment by Population and Contribution in European
Parliament After Brexit 109
   *Cesarino Bertini, Gianfranco Gambarelli, Izabella Stach,
and Giuliana Zibetti*

The Use of Group Decision-Making to Improve the Monitoring
of Air Quality 127
   *Cezary Orłowski, Piotr Cofta, Mariusz Wąsik, Piotr Welfler,
and Józef Pastuszka*

Bi-proportional Apportionments 146
   Mirko Bezzi, Gianfranco Gambarelli, and Giuliana Angela Zibetti

A Probabilistic Unified Approach for Power Indices in Simple Games 162
   Josep Freixas and Montserrat Pons

The Story of the *Poor* Public Good Index 171
    Manfred J. Holler

Author Index 181

A Probabilistic Model for Detecting Gerrymandering in Partially-Contested Multiparty Elections

Dariusz Stolicki[1,2(✉)], Wojciech Słomczyński[1,3], and Jarosław Flis[1,4]

[1] Jagiellonian Center for Quantitative Research in Political Science,
Jagiellonian University, ul. Wenecja 2, 31-117 Kraków, Poland
{dariusz.stolicki,jaroslaw.flis}@uj.edu.pl,
wojciech.slomczynski@im.uj.edu.pl
[2] Faculty of International and Political Studies,
Jagiellonian University, Kraków, Poland
[3] Faculty of Mathematics and Computer Science,
Jagiellonian University, Kraków, Poland
[4] Faculty of Management and Social Communication,
Jagiellonian University, Kraków, Poland

Abstract. Classic methods for detecting gerrymandering fail in multi-party partially-contested elections, such as the Polish local election of 2014. A new method for detecting electoral bias, based on the assumption that voting is a stochastic process described by Pólya's urn model, is devised to overcome these difficulties. Since the partially-contested character of the election makes it difficult to estimate parameters of the urn model, an ad-hoc procedure for estimating those parameters in a manner untainted by potential gerrymandering is proposed.

Keywords: Gerrymandering · Partially-contested elections · Pólya's urn model · Dirichlet distribution · Seats-votes relationship

1 Introduction

Most political representative bodies in the world are chosen through multi-district elections, where seats are apportioned among n parties within each of c districts independently, i.e., solely on the basis of the district vote. In such elections, jurisdiction-wide distribution of seats (*the seat distribution*) depends heavily not only on the overall voting result (i.e., a vector of party vote shares), but on the geographical distribution of each party's support over the set of electoral districts (*the vote distribution*). Anomalous vote distributions can lead to skewed electoral results, such as the well-known referendum paradox [11,49,59]. While

Supported by the Polish National Science Center (NCN) under grant no. 2014/13/B/HS5/00862, *Scale of gerrymandering in 2014 Polish township council elections.*

such anomalous distributions can arise through natural causes, such as voter self-segregation and other population clustering effects [18,38,42,43,75] (the U.S. electoral college, where two out of five most recent elections involved instances of the referendum paradox favoring the Republican Party, affords a prominent example), they can also be facilitated through deliberate manipulation of electoral district boundaries. Such manipulation, especially when undertaken for the purpose of obtaining an advantage for the party or block of parties controlling the redistricting process, is known as *gerrymandering*.

Gerrymandering is possible under all kinds of voting rules [6], but is most common under the combination of single-member electoral districts and the plurality rule (known in political science as the FPTP system). A classic gerrymander under FPTP is based on a combination of two strategies: assigning as many opposition voters as possible into a small number of districts (*packing*) (obviously, that number needs to be smaller than $c/2$), while spreading out the remainder roughly equally across other districts in such manner that they do not constitute a majority in any of them (*cracking*) [5,27]. When done correctly, this results in a substantial number of opposition votes in the "packed" districts being wasted, while the opposition supporters in other districts are so diluted that they are incapable of securing a plurality in any of them. If there are more than two parties, other strategies also become possible, such as *stacking*, balancing the number of supporters of different opposition parties in such manner that enables the preferred candidate to win with less than majority, but they tend to require more detailed knowledge about voter preferences and their distribution.

Ultimately, however, even as both strategies and objectives of gerrymandering are well-understood, the concept itself, as we will see below, remains difficult to formalize. Even apart from difficulties necessarily involved in discerning intent (and hence distinguishing manipulation from unintentional bias), there is no accepted standard by which a specific vote distribution can be judged "fair" or "natural" [16,36,39]. Without such standard, the concepts of distributional "unfairness" or "anomalousness" are fuzzy at best and meaningless at worst. This obviously makes it more difficult to detect and identify gerrymandering, as resort has to be had to circumstantial or otherwise indirect evidence.

2   Methodological Approaches to Detecting Gerrymandering

Altman et al. [2] distinguish six basic methodological approaches to detecting gerrymandering: *method of stated intent*, which relies on public statements of the authors of the districting plan; *method of totality of the circumstances*, which focuses on the political circumstances (well-known geographical rivalries, past practice, etc.); *method of evaluation of process*, which analyzes the districting process; *methods of inspection*, where gerrymandering is inferred from some qualitative or quantitative characteristics of the districting plan; *method of post-hoc comparisons*, where the districting plan is compared against a random sample of alternative plans; and *method of revealed preferences*, where the districting plan is compared against alternatives rejected during the districting process. Of those, the first three are purely qualitative and only rarely will suffice to prove gerrymandering, or even systemic bias. In addition, they require extensive extrinsic knowledge about the districting process that cannot be obtained from election results and districting plans alone. The method of revealed preferences, while advocated by [2], also requires such extrinsic knowledge (namely the set of plans that were known to the districting authority but have been rejected).

That leaves us with only two classes of quantitative methods for detecting gerrymandering that can be applied when extrinsic knowledge is unavailable: the methods of inspection and the method of post-hoc comparisons. As noted, in the former we focus on some observable characteristics of the districting plan and compare them against a well-known standard. Most of such methods focus on one of the basic types of plan characteristics: district geometry and the relation between seats and votes. Geometric methods involve tests of district contiguity and of various measures of district compactness [1, 28, 62, 77], trying to formalize the intuition that gerrymandered electoral districts are oddly shaped. Yet the evidence of manipulation provided by such methods is circumstantial at best, as irregularity of shape is neither necessary nor sufficient condition for gerrymandering.

Methods focusing on the seats-votes relation instead start from some assumptions about the desired characteristics of such relation. Such characteristics may include proportionality [8], responsiveness to shifts in voter support (measured by the swing ratio, i.e., the derivative of seats with respect to votes) [56, 73], partisan symmetry (a requirement that seats-votes curves by identical for all competing parties [32, 33, 35, 37, 47, 57]), or the efficiency principle, requiring that the number of wasted votes be equal for all parties [53, 69]. Then each party's seats-votes function (i.e., a function assigning to total vote share v the total seat share s) is tested for deviation from the chosen characteristics. Those methods generally share three principal limitations. The first one is of fundamental nature: most of the methods described above (except the partisan symmetry method) involve a priori assumptions that certain form of the seats-votes function is a natural one, but no attempt is made to justify those assumptions, for instance by showing that they arise from some general or accepted distributional assumptions. Without such justification it may well be that those methods generate a large number of false negatives by holding districting plans to a more restrictive standard than mere absence of distributional anomalies. The second problem with methods focusing on the seats-votes relation is more technical: they usually require that the full seats-votes function be known for each party, yet all that is empirically known is a single data point per election. Extrapolation from those data points involves questionable assumptions about how changes in one party's vote share translate into changes in its vote distribution and in other parties' vote shares (like the uniform partisan swing assumption, see [14, 15, 29, 35, 58]). Finally, virtually all methods focusing on the seats-votes relation have been developed with two-party elections in mind and usually lack natural generalization for multiparty elections.

The method of post-hoc comparisons instead compares districting plans with an ensemble of alternative districting plans [18–20, 24, 51, 60]. The problem is that the full set of correct solutions to the districting problem is in all but simplest cases too numerous to be used for such comparison, so we are reduced to testing the empirical plan against some sample of algorithmically generated random plans. But for proper inference to be drawn from such sample, we need the sample to be drawn from the set of all possible districting plans with some known probability measure, and we are unaware of any algorithm for generating districting plans for which such measure has been analytically determined [2,3].

Finally, all of the methods described above fail in *partially-contested elections*, i.e., those where only some parties (or even none) field candidates in all electoral districts, and other parties field candidates in fewer than all districts (including one-candidate parties that only run in a single district). In such cases, the number of candidates can vary across districts, affecting both vote distributions and seats-votes relationships. In addition, we are no longer free to generate alternative seat allocations by rearranging districts, since we no longer have data about each party's support beyond the districts it contested. It has been already noted by [45] that traditional statistical methods for dealing with missing data cannot be applied to partially-contested elections since failure to contest an election in a district is usually not a random event, but a function of the party's forecasted electoral strength in such district. Yet the methods for dealing with partially-contested elections proposed by, inter alia, [45,50,72,76], are also insufficient when the patterns of electoral contestation are very chaotic, and particularly if the election cannot be described as a mixture of relatively few patterns with multiple districts per each.

We have encountered exactly those problems when analyzing gerrymandering in Polish local election of 2014, which was held under the plurality rule. Due to highly personalized nature of local politics in Poland (especially in the smallest but most numerous class of municipalities, the townships), in 2386 out of 2412 municipalities the election was partially-contested. The chaotic character of electoral contestation patterns is best described by the following selection of facts:

- only 2218 out of 16,971 parties[1] have contested the election in all districts within their respective municipalities,
- if parties were ordered according to the fraction of districts contested within their respective municipalities, a median party would have contested less than half of all districts,
- 4733 parties have contested only a single district,
- there are, on average, 8.26 different district contestation patterns per municipality.

To address the problems described above, we propose a new method for detecting gerrymandering in partially-contested multiparty elections that are conducted under identical rules in multiple jurisdictions with separate districting plans (examples include regional and local elections, but also national elections

---

[1] Under the rules in place for the election, parties are registered at the district level, and every independent candidate is counted as a distinct party, thence the unusually large number of parties.

in which redistricting is done by subnational jurisdictions, as in the case of the U.S. House of Representatives). We proceed on two general assumptions: that voting in each district can be modeled by a stochastic process that is identical (modulo choice of parameters) for all jurisdictions of interest, and that gerrymandering is ultimately an exception rather than a rule, so the parameters of the stochastic model estimated from the set of all jurisdictions are free from the taint of manipulation. We first formulate a general model of vote distribution, then propose a procedure for estimating that model's parameters, and finally use that model to derive a sampling distribution of seat shares against which party seat shares can be compared.

3 Modeling District-Level Vote Distribution

3.1 Definitions and Notation

1. An electoral jurisdiction consists of a finite set of electoral districts D, whose cardinality we denote as $c := |D|$, a finite set of parties $P := \{1, \ldots, n\}$, and a left- and right-total relation $R \subseteq P \times D$ such that $(i, k) \in R$ if the i-th party fields a candidate in the k-th district. It is assumed here that in each district there is exactly one seat to be allocated using the plurality rule and hence each party is able to field only a single candidate.

2. Let $D_i \subseteq D$ be the set of indices of the electoral districts where the i-th party, $i \in P$, fields candidates, i.e., a set of such $k \in D$ that $(i, k) \in R$. Let $c_i := |D_i|$.

3. Let $P_k \subseteq P$ be the set of indices of the parties contesting the k-th district, $k \in D$, i.e., a set of such $i = 1, \ldots, n$ that $(i, k) \in R$. Let $n_k := |P_k|$.

4. Let $\sim$ be an equivalence relation on D identifying districts contested by the same set of parties, i.e., such that $k \sim l$ if and only if $P_k = P_l$. By $[k]_\sim$ we denote an equivalence class of k in D with respect to $\sim$. We call it a *contestation pattern*.

5. The *voting result* in the k-th district is a vector $\mathbf{v}_k := \left(v_{i_1}^k, \ldots, v_{i_{n_k}}^k \right) \in \delta_k$, where δ_k is an n_k-face of the standard $(n-1)$-dimensional simplex Δ_n that includes vertices i_1 to i_{n_k}, v_i^k is the *i-th party's vote share* in the k-th district, and $i_1 < \cdots < i_{n_k}$ are elements of P_k. Note that δ_k can be identified with the standard $(n_k - 1)$-dimensional simplex Δ_{n_k}.

6. Let $v_i := \left(\sum_{k \in D_i} v_i^k w_k \right) / \left(\sum_{k \in D_i} w_k \right)$, where w_k is the number of voters in the k-th district, be the i-th *party's total vote share*.

7. Let $\boldsymbol{D}_m$ be the set of all such districts $k \in \bigcup D$, where the sum is over all electoral jurisdictions of interest, that $n_k = m$.

8. By *quantile mixture* of absolutely continuous probability distributions $\mathcal{M}_1, \ldots, \mathcal{M}_m$ supported on some compact I we understand a probability distribution characterized uniquely by the inverse cumulative distribution function $\Lambda^{-1} : [0, 1] \to I$ given by $\Lambda^{-1}(x) := \frac{1}{m} \sum_{i=1}^m F_i^{-1}(x)$, where F_i^{-1} is the inverse cumulative distribution function of $\mathcal{M}_i$ [44].

9. Where single-district models are discussed (in Sects. 3.2 and 3.3) index k is omitted.

3.2   Overview of Available Models

The problem of modeling voter choice in single-choice electoral systems can be though of as a special case of the problem of modeling preference orderings, which is well known in the social choice theory (see, e.g., [65] and [71]). A number of models has been employed for that purpose, but, since we are only interested in the first choice, we omit the discussion of those that differ only in their treatment of the second and subsequent preferences.

1. Under the *Impartial Culture (IC)* model, each preference ordering (and, therefore, also each choice of the first candidate) is equiprobable and each voter decides independently with fixed probabilities [17]. The voting result **v** follows a multinomial distribution centered at the barycenter of Δ_n with the variance of the square distance from the barycenter $O\left(w^{-1}\right)$. There is extensive evidence for the claim that both the equiprobability and independence assumptions are not satisfied in empirical elections (recounted by, inter alia, [65]).

2. The *multinomial model* is a generalization of the IC model which assigns unequal probabilities to the candidates, but still assumes that each voter makes an independent choice with fixed probabilities described by vector **p**. The voting result **v** follows a multinomial distribution centered at **p** with the variance of the square distance from **p** behaving as $O\left(w^{-1}\right)$. As first noted in [46], this model significantly underestimates the variance of the vote distribution. To avoid that problem, Penrose and others [61, 67, 74] have proposed *clustered multinomial model*, according to which each district's population consists of κ equally sized clusters of voters who have identical characteristics and instead of randomizing individual voters' choices, we randomize each cluster's choice. Under that model, **v** still follows a multinomial distribution centered at **p**, but its variance increases to $O\left(\kappa^{-1}\right)$ (as $\kappa \ll w$ – Penrose's original estimate for Great Britain was $\kappa \approx 14$).

3. The *Impartial Anonymous Culture (IAC)* model treats each preference profile (and, therefore, each voting result) as equiprobable [31, 48]. Accordingly, the voting result **v** follows the uniform distribution on a discrete grid of points within Δ_{n_k}, which, as w approaches ∞, weakly converges to the uniform distribution on Δ_{n_k}.

4. The *Pólya urn model*, first introduced by Eggenberger and Pólya in 1923 [22], has been applied in the field social choice theory by, inter alia, [12, 21, 40, 66]. Voting is treated as a discrete stochastic process where a ball is drawn from an urn that initially contains α_i balls of the i-th color (where $i = 1, \ldots, n$), and after each draw λ balls of the same color as the one drawn are returned to the urn. The voting result **v** follows the multivariate Pólya distribution and, as w approaches ∞, converges almost surely to a random variable having the Dirichlet distribution parametrized by vector $(\alpha_1, \ldots, \alpha_n)/\lambda$ [7, 41]. Both IC and IAC are special cases of the urn model, with $\alpha_1 = \cdots = \alpha_n = 1$ and $\lambda = 1$ for IAC and 0 for IC.

5. *Spatial models* assume that voter policy preferences are distributed (usually normally) over a multidimensional policy space, that party policy positions are either specified or randomly distributed over the same space and that voters always choose the candidate of the closest party according to some fixed metric [4, 23, 55]. Again, some voting clustering has to be assumed to avoid overestimating homogeneity.

There is considerable evidence that equiprobability and independence assumptions fail to match empirical data, and accordingly both IC and IAC fail as empirical models of electoral behavior [65, 71]. In [71], spatial models are found to be most effective in modeling preference profiles, but in single-choice elections such models involve too many degrees of freedom for estimation unless highly simplifying assumptions are made (for instance, about reduction of the number of dimensions). That leaves only the urn model for our intended applications.

Sociological theory of electoral behavior also provides sound reasons for adopting the urn model. Contagion mechanisms it is used to model translate into an observation that most voters are initially undecided and their political views are shaped through social interactions with others, who include already-committed supporters of the parties and candidates (cf. [13]). Indeed, political parties recognize that direct mobilization of voters through personal interaction is one of the most important tools of electoral campaigning [26]. Even mass media influence on political views, which would seem to support rather fixed-probability models, is indirect and effective primarily when the information communicated by the media is later verified through direct interaction with other members of the community [54]. It is also recognized that such political contagion processes are essentially stochastic, being dependent on the fine structure of social networks [52] which cannot be predicted deterministically.

3.3 Urn Model of Electoral Behavior

A Pólya urn model is usually characterized by two parameters: a vector of initial ball numbers ($\alpha \in \mathbb{R}_+^n$) and the number of additional balls returned after each draw ($\lambda \in \mathbb{R}_+ \cup \{0\}$), but note that by rescaling vector α we can always obtain $\lambda = 1$, thereby reducing our parameter space to $\mathbb{R}_+^n$. In addition, it is often convenient to express α as a product of an n-element vector $\mathbf{p} \in \Delta_n$ and of the concentration parameter $\alpha \in \mathbb{R}_+$.

Definition 1. *Pólya-Eggenberger Urn Model [22, 63].*

Let us consider a countably infinite set of potential voters. Let $X_j \in P$ be the choice of the j-th voter ($j \in \mathbb{N}$, $\mathbb{N} = \{1, 2, 3, \dots\}$).

Voting is a discrete stochastic process where the probability of the $(j + 1)$-th voter choosing the i-th party's candidate is defined by induction as

$$\Pr\left(X_{j+1} = i\right) = \frac{\alpha p_i + |\{k = 1, \dots, j : X_k = i\}|}{\alpha + j}, \tag{1}$$

for $i = 1, \dots, n$.

Intuitively, the attractiveness of the i-th party to the $(j+1)$-th voter is proportional to the sum of the number of voters that already have decided to support it and its initial strength αp_i. In [66] the authors propose that αp_i be interpreted as the number of voters who are committed at the outset to support the i-th party's candidate, but this interpretation raises some issues as αp_i need not be an integer.

Proposition 1. *In the above situation, there exists a random variable* $\mathbf{V} \sim$ Dir $(n; \alpha p_1, \ldots, \alpha p_n)$ *such that* $(\Pr(X_j = 1), \ldots, \Pr(X_j = n)) \overset{a.s.}{\to} \mathbf{V}$ *as* $j \to \infty$, *where* Dir $(n; \alpha p_1, \ldots, \alpha p_n)$ *(the Dirichlet distribution) is a continuous multivariate probability distribution supported on* Δ_n *that has a probability density* f *with respect to the Lebesgue measure on* Δ_n *given by:*

$$f(v_1, \ldots, v_n) := \frac{1}{\mathrm{B}(\alpha_1, \ldots, \alpha_n)} \prod_{i=1}^{n} v_i^{\alpha_i - 1}, \tag{2}$$

where $\mathbf{v} \in \Delta_n$ *and* $\mathrm{B}(\alpha_1, \ldots, \alpha_n)$ *is the multivariate beta function:*

$$\mathrm{B}(\alpha_1, \ldots, \alpha_n) := \frac{\prod_{i=1}^{n} \Gamma(\alpha_i)}{\Gamma(\alpha)}. \tag{3}$$

In the above situation we have for $i = 1, \ldots, n$:

$$V_i \sim \text{Beta}(\alpha p_i, \alpha(1 - p_i)); \tag{4}$$

$$\mathrm{E}(V_i) = p_i; \tag{5}$$

$$\mathrm{Var}(V_i) = \frac{p_i(1 - p_i)}{\alpha + 1}. \tag{6}$$

For proof of the above proposition see, inter alia, [7] and [41] (Fig. 1).

3.4   Parameter Fitting – The Expectation Vector

Literature on electoral studies recognizes that district-level vote shares depend on two principal factors: overall party popularity, measured by the total vote share vector $\mathbf{v}$, and political geography, i.e., district-specific effects, which are more difficult to model formally. However, as we consider an idealized distribution of vote shares in a non-biased election, in essence approximating an average distribution of district vote shares over the population of non-biased districting plans, we abstract from the effects of political geography altogether.

It would thus appear from (5) that the vector of party total vote shares $\mathbf{v}$ would be the most natural estimate of parameter $\mathbf{p}$. This, however, is only the case if the voting results in all districts in D come from a single distribution, which in turn is equivalent to a condition that the election be fully contested, i.e., that every party j field a candidate in every district k. Otherwise, there must be a different distribution for each equivalence class $[k]_\sim$, as each such class is characterized by the presence of a different set of parties. It follows that in

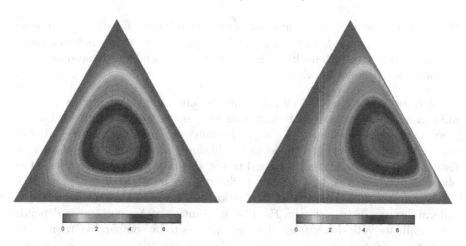

Fig. 1. Density plot of a symmetric Dirichlet distribution on Δ_3 with $\mathbf{p} = (\frac{1}{3}, \frac{1}{3}, \frac{1}{3})$ and $\alpha = 9$ (left) and of an asymmetric Dirichlet distribution on Δ_3 with $\mathbf{p} = (\frac{4}{9}, \frac{1}{3}, \frac{2}{9})$ and $\alpha = 8$ (right).

partially-contested elections, which are of primary interest to us, voting results in D will be distributed according to a direct product of Dirichlet distributions $\mathcal{D}_k := \text{Dir}(n_k; \alpha_k \mathbf{p}_k)$, with n_k, $\mathbf{p}_k$ and α_k constant for each equivalence class $[k]_\sim$, and $\mathbf{p}_k$ and α_k being unknown.

We cannot simply assume that $p_i^k = \sum_{j \in [k]_\sim} v_i^j / |[k]_\sim|$ for each $k \in D$, since the empirical vote distribution over equivalence classes $[k]_\sim$ may be tainted by gerrymandering. Instead, we need a theoretical model that is based solely on district contestation patterns (described by relation R) and party total vote shares vector $\mathbf{v}$.

In fitting $\mathbf{p}_k$ to each equivalence class in D with respect to $\sim$, we seek to satisfy the following three natural requirements:

R1 For each district k, $\mathbf{p}_k \in \Delta_{n_k}$, i.e., $\sum_{i \in P_k} p_i^k = 1$.

R2 For any two districts $k, l \in D$, if $k \sim l$, then $\mathbf{p}_k = \mathbf{p}_l$.

R3 For any two parties $i, j \in P$, the order on $\{p_i^k, p_j^k\}$ is identical in every district $k \in D_i \cap D_j$.

R4 For any two parties $i, j \in P$ such that $D_i = D_j$, the order on $\{p_i^k, p_j^k\}$ is identical with the order on $\{v_i, v_j\}$ for each district $k \in D_i$.

In addition, there are three postulates that we seek to satisfy approximately (i.e., to minimize deviation from them):

P5 For each party $i \in P$ its mean expected vote share over districts should be close to its party vote share, i.e., $\sum_{k \in D_i} p_i^k \approx c_i v_i$.

P6 For any two districts $k, l \in D$ if $n_k = n_l$, then $p_i^k \approx p_i^l$ for each party $i \in P_k \cap P_l$.

P7 For each party $j \in P$ and for any two districts $k, l \in D_j$ we have $\varphi_{n_k}\left(p_j^k\right) = \varphi_{n_l}\left(p_j^l\right)$, where $\varphi_m : [0,1] \to [0,1]$, $m \in \mathbb{N}$, is a function mapping a party vote share in a district with m contenders to a standardized value independent of m.

Of those postulates, **P7** clearly requires some additional discussion. The underlying problem consists of comparing vote shares across districts with different number of candidates. Clearly, obtaining 40% of the vote in a district with two candidates is not equivalent to obtaining an identical vote share in a district with ten candidates. In formal terms, this intuition can be expressed as follows: let X_m, $m \in \mathbb{N}$, be a random variable given by $X_m(i,k) := v_i^k$, where k is drawn from a uniform discrete distribution on D_m and i is later drawn from a uniform discrete distribution on P_k. The distribution of X_m necessarily depends on m, while for vote shares from different districts to be comparable, we need to transform X_m into another random variable with a distribution that is constant with respect to m.

The probability integral transform of X_m is one natural choice of such transformation. Let us consider the cumulative distribution function of X_m. As it is not injective, X_m being discrete, we formally define $\varphi_m : [0,1] \to [0,1]$ as its continuous approximation obtained by integrating the probit-transformed [2] kernel density estimator ψ_m [30] of the distribution of X_m, i.e., $\varphi_m(p) = \int_0^p \psi_m(x)\,dx$ for $p \in [0,1]$. This assures that φ_m is invertible, and that φ_m^{-1} is continuous, strictly increasing, and the images of the bounds of its domain are, respectively, 0 and 1. It follows that every linear combination of functions φ_k^{-1}, where $k \in \mathbb{N}$, with positive coefficients summing up to $c > 0$, is also continuous and strictly increasing, and the images of the bounds of its domain are 0 and c. Let $i \in P$. By the intermediate value theorem there exists a unique $q_i \in [0,1]$ such that $\sum_{k \in D_i} \varphi_{n_k}^{-1}(q_i) = c_i v_i \le c_i$. Hence the definition $p_i^k := \varphi_{n_k}^{-1}(q_i)$ would naturally imply **P5**. Parameter q_i has no natural interpretation, however if we assume the distribution of the i-th party's district vote shares to be a quantile mixture of the distributions $\mathcal{D}_k$, where $k \in D_i$, q_i will correspond to the value of such mixture's cumulative distribution function Λ_i for the empirical value of v_i.

Note that model assuming $p_i^k = \varphi_{n_k}^{-1}(q_i)$ satisfies most of the requirements and postulates specified above:

- **P5** and **P7** are satisfied by definition of q_i and φ_m.
- If $n_k = n_l$, $p_i^k = \varphi_{n_k}^{-1}(q_i) = \varphi_{n_l}^{-1}(q_i) = p_i^l$ for any party i and any two districts $k, l \in D_i$, so **P6** is satisfied exactly and therefore implies **R2**.
- **R3** results from the monotonicity of φ_m^{-1}.
- From the monotonicity of φ_m^{-1} we know that the order on $\{p_i^k, p_j^k\}$ is identical with the order on $\{q_i, q_j\} = \{\Lambda_i(v_i), \Lambda_j(v_j)\}$. From $D_i = D_j$ it follows that $\Lambda_i = \Lambda_j$. As Λ_i is strictly increasing, the order on $\{q_i, q_j\}$ is identical to that on $\{v_i, v_j\}$, as desired under **R4**.

---

[2] We use the probit-transformed kernel density estimator instead of a standard Gaussian density estimation to ensure that the resulting estimator is of bounded support and that φ_m is surjective onto $[0,1]$.

Unfortunately, there is no guarantee that the above model satisfies **R1**. We therefore modify it by renormalizing vector $\mathbf{p}_k$ for each district k. This renormalization ensures that **R1** is satisfied, **R3** is preserved (as renormalization preserves the ordering of $p_1^k, \ldots, p_n^k$), and so are **R2** and **R4** (as the renormalization constant does not vary within $[k]_\sim$). In turn, such renormalization may introduce violations of **P5**, **P6**, and **P7**, but we do not need those postulates to be satisfied exactly.

Note that this method is loosely analogous to the biproportionality method by [9,10,64].

The distribution of p_i^k in all districts $k \in \boldsymbol{D}_m$, $m \in \mathbb{N}$, will be of further interest in the following section (recall that $\boldsymbol{D}_m$ is the set of all districts with m candidates). We denote its cumulative distribution function by Ψ_m.

3.5 Parameter Fitting – The Concentration Parameter

The last parameter of our electoral model is the concentration parameter α_k. Unlike the expected vote shares of the contending parties, α_k is never observable directly, and in most cases we do not have enough data to fit it to empirical voting results using some distribution fitting method that produces a reasonable confidence interval (since such fitting would require large equivalence class $[k]_\sim$). Intuitively, the concentration parameter should depend on at least two further parameters: the number of candidates and the political homogeneity of the jurisdiction under consideration. The latter, in turn, is likely to depend in a complex manner on a large number of factors, such as the population and area of the jurisdiction, historical cleavages, settlement structure, socioeconomic diversity, etc. We do not have a good theoretical model of those relationships that would enable us to estimate α_k and a formulation of such model would go far beyond the scope of this paper.

To circumvent this issue we treat the concentration parameter as another random variable distributed, for each class of districts $\boldsymbol{D}_m$, with a gamma distribution with parameters κ_m and θ_m. To apply our model to a particular class of elections, we still need to estimate those parameters of the distribution of the concentration parameter. We proceed as follows: let Y_m, $m \in \mathbb{N}$, be a random variable given by $Y_m(i,k) := V_i^k$, where V_i^k is the i-th barycenteric coordinate of $\mathbf{V}_k \sim \mathrm{Dir}(n_k; \alpha_k \mathbf{p}_k)$, k is drawn from a uniform discrete distribution on $\boldsymbol{D}_m$ and i is later drawn from a uniform discrete distribution on P_k. Intuitively, it is the theoretical vote share of a random party in a random district in an ideal unbiased election. Under our model, the distribution of Y_m is a compound beta distribution with parameters $(\alpha p, \alpha - \alpha p)$ (see (4)), where $p \sim \Psi_m$ and $\alpha \sim \mathrm{Gamma}(\kappa_m, \theta_m)$. Accordingly, the density of that distribution is given by:

$$f_m(x) = \int_0^1 \int_0^\infty \frac{x^{\alpha p - 1}(1-x)^{\alpha - \alpha p - 1}}{\mathrm{B}(\alpha p, \alpha - \alpha p)} \frac{\alpha^{\kappa_m - 1} e^{-\frac{\alpha}{\theta_m}}}{\Gamma(\kappa_m)\theta_m^{\kappa_m}} \, d\alpha \, d\Psi_m(p). \tag{7}$$

The function Φ_m is known at this stage (having been estimated in the preceding section), so the only two unknowns in this formula are the gamma distribution parameters κ_m and θ_m. But note that under our general assumption that gerrymandering is not ubiquitous, the distribution of Y_m should closely approximate the distribution of X_m provided the model is correct. Therefore, we can use that property to obtain κ_m and θ_m. We do that by numerically minimizing, for each $m \in \mathbb{N}$, the total variation distance [34,68] between the distributions of X_m and Y_m.

4   Modeling the Sampling Distribution of Seats

By this point, we have estimated all the parameters necessary to model the ideal unbiased distribution of votes in each electoral district in every jurisdiction, namely, n_k and $\mathbf{p}_k$ for each district k, and κ_m and θ_m for each class of districts D_m, $m \in \mathbb{N}$. Of course, not all anomalies in the vote distribution of a party are of interest to us when seeking to detect gerrymanders, but only those that translate into biases in the allocation of seats. To detect such biases, we run a Monte Carlo simulation for each jurisdiction of interest, generating a large number of simulated election results. We proceed as follows:

1. For each district $k \in D$, we generate a single realization of the random variable $\alpha_k \sim \text{Gamma}\,(\kappa_{n_k}, \theta_{n_k})$, which we will denote as $\widehat{\alpha}_k$.
2. For each district $k \in D$, we then generate a single realization of the random vector $\mathbf{V}_k \sim \text{Dir}\,(n_k; \widehat{\alpha}_k \mathbf{p}_k)$, which we will denote as $\widehat{\mathbf{V}}_k$.
3. We distribute seats within each district $k \in D$ according to the plurality rule, awarding a single seat to the party with the greatest vote share, i.e., to the one corresponding to the greatest barycentric coordinate of $\widehat{\mathbf{V}}_k$.
4. For each party $i \in P$ we sum seats over districts.

This procedure is repeated 2^{20} times for each electoral jurisdiction. Through this process, we obtain a joint discrete sampling distribution of party seat vectors $\mathcal{S}$ on $\prod_{i=1}^{n}\{0, \ldots, c_i\}$, and for each party $i \in P$ we denote the marginal sampling distribution of seats by $\mathcal{S}_i$. In the process of estimating the above distributions we do not rely on the empirical distribution of voters among districts, and therefore they are untainted by the possible gerrymandering.

To measure the distance between actual seat count of the i-th party, s_i, and the distribution $\mathcal{S}_i$ obtained above, we introduce a simple measure analogous to the well-known p-value used in statistical hypothesis testing:

$$\pi_i := \min\left(\mathcal{S}_i\left([0, s_i]\right), \mathcal{S}_i\left([s_i, c_i]\right)\right). \tag{8}$$

In other words, π_i is the probability of a party obtaining the number of seats that is equal to or more extreme than its actual number of seats. Low values of π_i are indicative not only of anomalies in the vote distribution, but also of the fact that they translate into a rather improbable deviation from the expected number of seats.

To obtain a jurisdiction-level index, we could simply average the values of π_i over $i \in P$. However, to account for the fact that we are primarily interested in cases of gerrymandering affecting parties contesting most districts, we weigh π_i by the number of districts c_i. The resulting index,

$$\pi := \frac{\sum_{i=1}^{n} \pi_i c_i}{\sum_{i=1}^{n} c_i}, \tag{9}$$

is our final measure of electoral bias. While not conclusive evidence of gerrymandering, since we still lack proof of intent, as electoral bias can be unintentional and arise due to pecularities of spatial distribution of party voters), it allows us to identify the outlier jurisdictions which can then be analyzed using other, possibly more qualitative methods.

Remark 1. Note that for our primary data set of interest, Polish local elections of 2014, π is quite well approximated by a normal distribution, see Fig. 2. This indicates an absence of pervasive gerrymandering, which agrees with the intuition that gerrymandering (or at least successful gerrymandering) is more difficult in less orderly party systems.

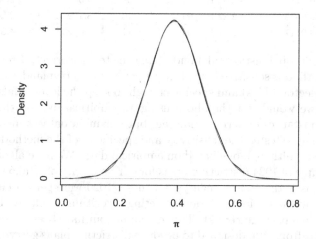

Fig. 2. A kernel density estimate of the empirical density of π for Polish local election of 2014 (black) and a normal density curve with $\mu \approx 0.388$ and $\sigma \approx 0.0945$ (red). (Color figure online)

To conclude, we have seen that classic methods for detecting gerrymandering fail when applied to multiparty partially-contested elections. We propose an alternative method based on a probabilistic model of voting behavior, together with a procedure for estimating the parameters of such model in a manner insulated from the possible taint of gerrymandering. We admit that the method is complex and involves simplifying assumptions and ad-hoc solutions, but they are made necessary due to the complexity of the problem and the limitations of the available data. Ultimately, we are unable to secure any conclusive evidence of gerrymandering, but we do obtain a single index that can be used to identify the suspect jurisdictions for further analysis.

A   Appendix

We have developed two proof-of-concept tests to evaluate the correctness of the proposed method for detecting gerrymandering. First, for any set of jurisdictions to which we seek to apply the proposed method we can test whether the empirical marginal distribution of X_m, $m = 1, \ldots, 11$, agrees with the theoretical marginal distribution expressed by (7). We have run such test for our primary data set of interest, Polish local election of 2014, producing, for the following values of parameters κ_m and θ_m (fitted in accordance with the procedure described in Sect. 3.5), the following total variation distances d_{TV} between the theoretical density function and the kernel density estimator of the empirical density (Fig. 3):

m	κ_m	θ_m	d_{TV}	m	κ_m	θ_m	d_{TV}
2	10.9656	1	4.97 E − 04	7	15.1239	1	2.51 E − 03
3	9.7069	1	1.42 E − 03	8	16.4486	1	2.73 E − 03
4	10.8787	1	2.02 E − 03	9	18.3641	1	2.93 E − 03
5	12.5535	1	2.22 E − 03	10	18.1292	1	3.24 E − 03
6	13.9117	1	2.27 E − 03	11	24.6423	1	4.33 E − 03

Even if the probabilistic model underlying the test is correct, it remains to be seen if the method is sensitive enough to detect actual gerrymandering (or other instances of electoral bias) and specific enough to keep the level of false positives low. Ideally, we would test the above using an empirical dataset that includes some known instances of gerrymandering, but our main dataset included none. Therefore, we have tested the sensitivity and specificity of the method against an artifical dataset, although one based on empirical data. We have algorithmically created a sample of 1024 districting plans for our home city of Kraków, each having 43 seats (as is the case in reality). Of those, 1020 were generated randomly using a Markov chain Monte Carlo districting algorithm developed by [25] and implemented in R package redist. The remaining four have been generated using two algorithms from [70], designed to produce districting plans gerrymandered in favor of one of the two largest parties (the first algorithm has also been designed to try to keep the districts relatively compact, while the second has been freed of all compactness constraints). Under each of those districting plans, we have calculated simulated election results using the 2014 precinct-level data. We treat each of such simulated elections as a single jurisdiction for which we carry out the procedure described in the article to obtain π. There are 12 distinct outcomes arising in simulated elections. As under all 1024 plans, all seats are won by the two largest parties, those outcomes are uniquely characterized by s_1 (or s_2).

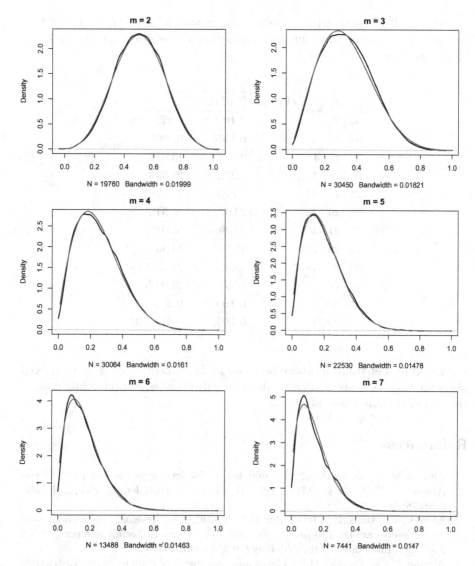

Fig. 3. The theoretical (red) and empirical (black) densities of v_i^k for different values of m. The empirical density is a kernel density estimate with the number of points and the bandwidth given below each plot. (Color figure online)

We list all of them, with corresponding π values, in the table below (the outcomes arising under the four intentionally gerrymandered plans are identified by the bold font):

s_1	s_2	No. of plans	% of plans	π
36	**7**	1	0.10%	**7.8E − 06**
31	**12**	1	0.10%	**0.39%**
26	17	1	0.10%	13.67%
25	18	3	0.29%	21.53%
24	19	43	4.20%	31.47%
23	20	226	22.07%	42.94%
22	21	433	42.29%	55.04%
21	22	262	25.59%	44.96%
20	23	48	4.69%	33.32%
19	24	4	0.39%	23.09%
13	**30**	1	0.10%	**0.48%**
6	**37**	1	0.10%	**4.4E − 07**

As can be seen from the above table, for all four intentionally gerrymandered plans the value of π are small enough to identify them as suspect, while none of the unbiased plans are so identified.

References

1. Altman, M.: Modeling the effect of mandatory district compactness on partisan gerrymanders. Polit. Geogr. **17**(8), 989–1012 (1998). https://doi.org/10.1016/S0962-6298(98)00015-8
2. Altman, M., Amos, B., McDonald, M.P., Smith, D.A.: Revealing preferences: why gerrymanders are hard to prove, and what to do about it. Technical report, SSRN 2583528, March 2015. https://doi.org/10.2139/ssrn.2583528
3. Altman, M., McDonald, M.P.: The promise and perils of computers in redistricting. Duke J. Const. Law Public Policy **5**(1), 69–159 (2010)
4. Ansolabehere, S., Leblanc, W.: A spatial model of the relationship between seats and votes. Math. Comput. Model. **48**(9–10), 1409–1420 (2008). https://doi.org/10.1016/j.mcm.2008.05.028
5. Apollonio, N., Becker, R.I., Lari, I., Ricca, F., Simeone, B.: The sunfish against the octopus: opposing compactness to gerrymandering. In: Simeone, B., Pukelsheim, F. (eds.) Mathematics and Democracy: Recent Advances in Voting Systems and Collective Choice, pp. 19–41. Springer, Berlin-Heidelberg (2006). https://doi.org/10.1007/3-540-35605-3_2
6. Aras, A., Costantini, M., van Erkelens, D., Nieuweboer, I.: Gerrymandering in three-party elections under various voting rules (2017)

7. Athreya, K.B.: On a characteristic property of Polya's urn. Stud. Sci. Math. Hung. **4**, 31–35 (1969)
8. Bachrach, Y., Lev, O., Lewenberg, Y., Zick, Y.: Misrepresentation in district voting. In: Proceedings of the Twenty-Fifth International Joint Conference on Artificial Intelligence, pp. 81–87 (2016)
9. Balinski, M.L., Demange, G.: Algorithms for proportional matrices in reals and integers. Math. Program. **45**(1–3), 193–210 (1989). https://doi.org/10.1007/BF01589103
10. Balinski, M.L., Demange, G.: An axiomatic approach to proportionality between matrices. Math. Oper. Res. **14**(4), 700–719 (1989). https://doi.org/10.1287/moor.14.4.700
11. Barthélémy, F., Martin, M., Piggins, A.: The architecture of the Electoral College, the house size effect, and the referendum paradox. Electoral. Stud. **34**, 111–118 (2014). https://doi.org/10.1016/j.electstud.2013.07.004
12. Berg, S.: Paradox of voting under an urn model: the effect of homogeneity. Public Choice **47**(2), 377–387 (1985). https://doi.org/10.1007/BF00127533
13. Blais, A.: Turnout in elections. In: Dalton, R.E., Klingemann, H.D. (eds.) Oxford Handbook of Political Behavior, pp. 621–635. Oxford University Press, Oxford (2007). https://doi.org/10.1093/oxfordhb/9780199270125.003.0033
14. Brookes, R.H.: Electoral distortion in New Zealand. Aust. J. Polit. Hist. **5**(2), 218–223 (1959). https://doi.org/10.1111/j.1467-8497.1959.tb01197.x
15. Butler, D.E.: Appendix. In: Nicholas, H.G. (ed.) The British General Election of 1950, pp. 306–333. Macmillan, London (1951)
16. Cain, B.E.: Assessing the partisan effects of redistricting. Am. Polit. Sci. Rev. **79**(02), 320–333 (1985). https://doi.org/10.2307/1956652
17. Campbell, C.D., Tullock, G.: A measure of the importance of cyclical majorities. Econ. J. **75**(300), 853 (1965). https://doi.org/10.2307/2229705
18. Chen, J., Rodden, J.: Unintentional gerrymandering: political geography and electoral bias in legislatures. Q. J. Polit. Sci. **8**(3), 239–269 (2013). https://doi.org/10.1561/100.00012033
19. Chen, J., Rodden, J.: Cutting through the thicket: redistricting simulations and the detection of partisan gerrymanders. Election Law J. Rules Politics Policy **14**(4), 331–345 (2015). https://doi.org/10.1089/elj.2015.0317
20. Cirincione, C., Darling, T.A., O'Rourke, T.G.: Assessing South Carolina's 1990s congressional districting. Polit. Geogr. **19**(2), 189–211 (2000). https://doi.org/10.1016/S0962-6298(99)00047-5
21. Coleman, J.S.: Introduction to Mathematical Sociology. Free Press, New York (1964)
22. Eggenberger, F., Pólya, G.: über die Statistik verketteter Vorgänge. Zeitschrift für Angewandte Mathematik und Mechanik **3**(4), 279–289 (1923). https://doi.org/10.1002/zamm.19230030407
23. Enelow, J.M., Hinich, M.J.: The Spatial Theory of Voting: An Introduction. Cambridge University Press, Cambridge (1984)
24. Engstrom, R.L., Wildgen, J.K.: Pruning thorns from the thicket: an empirical test of the existence of racial gerrymandering. Legis. Stud. Q. **2**(4), 465–479 (1977). https://doi.org/10.2307/439420
25. Fifield, B., Higgins, M., Imai, K.: A new automated redistricting simulator using Markov Chain Monte Carlo. Working Paper, Princeton University, Princeton, NJ (2015)

26. Foos, F., John, P.: Parties are no civic charities: voter contact and the changing partisan composition of the electorate. Polit. Sci. Res. Methods **6**(2), 283–298 (2018). https://doi.org/10.1017/psrm.2016.48

27. Friedman, J.N., Holden, R.T.: Optimal gerrymandering: sometimes pack, but never crack. Am. Econ. Rev. **98**(1), 113–144 (2008). https://doi.org/10.1257/aer.98.1.113

28. Fryer, R.G., Holden, R.: Measuring the compactness of political districting plans. J. Law Econ. **54**(3), 493–535 (2011). https://doi.org/10.1086/661511

29. Garand, J.C., Parent, T.W.: Representation, swing, and bias in U.S. Presidential Elections, 1872–1988. Am. J. Polit. Sci. **35**(4), 1011 (1991). https://doi.org/10.2307/2111504

30. Geenens, G.: Probit transformation for kernel density estimation on the unit interval. J. Am. Stat. Assoc. **109**(505), 346–358 (2014). https://doi.org/10.1080/01621459.2013.842173

31. Gehrlein, W.V., Fishburn, P.C.: Condorcet's paradox and anonymous preference profiles. Public Choice **26**(1), 1–18 (1976). https://doi.org/10.1007/BF01725789

32. Gelman, A., King, G.: Estimating the electoral consequences of legislative redistricting. J. Am. Stat. Assoc. **85**(410), 274–282 (1990)

33. Gelman, A., King, G.: A unified method of evaluating electoral systems and redistricting plans. Am. J. Polit. Sci. **38**(2), 514–554 (1994)

34. Gibbs, A.L., Su, F.E.: On choosing and bounding probability metrics. Int. Stat. Rev. **70**(3), 419–435 (2002). https://doi.org/10.1111/j.1751-5823.2002.tb00178.x

35. Grofman, B.: Measures of bias and proportionality in seats-votes relationships. Polit. Methodol. **9**(3), 295–327 (1983)

36. Grofman, B.: Criteria for districting: a social science perspective. UCLA Law Rev. **33**(1), 77–184 (1985)

37. Grofman, B., King, G.: The future of partisan symmetry as a judicial test for partisan gerrymandering after LULAC v. Perry. Election Law J. Rules Polit. Policy **6**(1), 2–35 (2007). https://doi.org/10.1089/elj.2006.6002

38. Gudgin, G., Taylor, P.J.: Seats, Votes, and the Spatial Organisation of Elections. Pion, London (1979)

39. Issacharoff, S.: Gerrymandering and political cartels. Harvard Law Rev. **116**(2), 593–648 (2002). https://doi.org/10.2307/1342611

40. Jamison, D., Luce, E.: Social homogeneity and the probability of intransitive majority rule. J. Econ. Theory **5**(1), 79–87 (1972)

41. Johnson, N.L., Kotz, S.: Urn Models and Their Applications: An Approach to Modern Discrete Probability Theory. Wiley, New York (1977)

42. Johnston, R.: Manipulating maps and winning elections: measuring the impact of malapportionment and gerrymandering. Polit. Geogr. **21**(1), 1–31 (2002). https://doi.org/10.1016/S0962-6298(01)00070-1

43. Johnston, R.J.: Political, Electoral, and Spatial Systems: An Essay in Political Geography. Contemporary Problems in Geography. Clarendon Press; Oxford University Press, Oxford; New York (1979)

44. Karvanen, J.: Estimation of quantile mixtures via L-moments and trimmed L-moments. Comput. Stat. Data Anal. **51**(2), 947–959 (2006). https://doi.org/10.1016/j.csda.2005.09.014

45. Katz, J.N., King, G.: A statistical model for multiparty electoral data. Am. Polit. Sci. Rev. **93**(1), 15–32 (1999). https://doi.org/10.2307/2585758

46. Kendall, M.G., Stuart, A.: The law of the cubic proportion in election results. Br. J. Sociol. **1**(3), 183–196 (1950). https://doi.org/10.2307/588113

47. King, G., Browning, R.X.: Democratic representation and partisan bias in Congressional elections. Am. Polit. Sci. Rev. **81**(4), 1251 (1987). https://doi.org/10.2307/1962588

48. Kuga, K., Nagatani, H.: Voter antagonism and the paradox of voting. Econometrica **42**(6), 1045–1067 (1974). https://doi.org/10.2307/1914217

49. Lepelley, D., Merlin, V., Rouet, J.L.: Three ways to compute accurately the probability of the referendum paradox. Math. Soc. Sci. **62**(1), 28–33 (2011). https://doi.org/10.1016/j.mathsocsci.2011.04.006

50. Linzer, D.A.: The relationship between seats and votes in multiparty systems. Polit. Anal. **20**(3), 400–416 (2012). https://doi.org/10.1093/pan/mps017

51. Mattingly, J.C., Vaughn, C.: Redistricting and the will of the people. Technical report, arXiv: 1410.8796 [physics.soc-ph], October 2014

52. McClurg, S.D.: The electoral relevance of political talk: examining disagreement and expertise effects in social networks on political participation. Am. J. Polit. Sci. **50**(3), 737–754 (2006). https://doi.org/10.1111/j.1540-5907.2006.00213.x

53. McGhee, E.: Measuring partisan bias in single-member district electoral systems: measuring partisan bias. Legis. Stud. Q. **39**(1), 55–85 (2014). https://doi.org/10.1111/lsq.12033

54. Mcleod, J.M., Scheufele, D.A., Moy, P.: Community, communication, and participation: the role of mass media and interpersonal discussion in local political participation. Polit. Commun. **16**(3), 315–336 (1999). https://doi.org/10.1080/105846099198659

55. Merrill, S.: A comparison of efficiency of multicandidate electoral systems. Am. J. Polit. Sci. **28**(1), 23–48 (1984). https://doi.org/10.2307/2110786

56. Niemi, R.G.: Relationship between votes and seats: the ultimate question in political gerrymandering. UCLA Law Rev. **33**, 185–212 (1985)

57. Niemi, R.G., Deegan, J.: A theory of political districting. Am. Polit. Sci. Rev. **72**(4), 1304–1323 (1978). https://doi.org/10.2307/1954541

58. Niemi, R.G., Fett, P.: The swing ratio: an explanation and an assessment. Legis. Stud. Q. **11**(1), 75–90 (1986). https://doi.org/10.2307/439910

59. Nurmi, H.: Voting Paradoxes and How to Deal with Them. Springer, New York (1999). https://doi.org/10.1007/978-3-662-03782-9

60. O'Loughlin, J.: The identification and evaluation of racial gerrymandering. Ann. Assoc. Am. Geogr. **72**(2), 165–184 (1982). https://doi.org/10.1111/j.1467-8306.1982.tb01817.x

61. Penrose, L.S.: On the Objective Study of Crowd Behaviour. H.K. Lewis, London (1952)

62. Polsby, D.D., Popper, R.D.: The third criterion: compactness as a procedural safeguard against partisan gerrymandering. Yale Law Policy Rev. **9**(2), 301–353 (1991)

63. Pólya, G.: Sur quelques points de la théorie des probabilités. Ann. l'inst. Henri Poincaré **1**(2), 117–161 (1930)

64. Pukelsheim, F.: Biproportional scaling of matrices and the iterative proportional fitting procedure. Ann. Oper. Res. **215**(1), 269–283 (2014). https://doi.org/10.1007/s10479-013-1468-3

65. Regenwetter, M., Grofman, B., Tsetlin, I., Marley, A.: Behavioral Social Choice: Probabilistic Models, Statistical Inference, and Applications. Cambridge University Press, Cambridge (2006)

66. Sano, F., Hisakado, M., Mori, S.: Mean field voter model of election to the House of Representatives in Japan. In: APEC-SSS2016, p. 011016. Journal of the Physical Society of Japan (2017). https://doi.org/10.7566/JPSCP.16.011016

67. Stanton, R.G.: A result of Macmahon on electoral predictions. Ann. Discrete Math. **8**, 163–167 (1980). https://doi.org/10.1016/S0167-5060(08)70866-5

68. Steerneman, T.: On the total variation and hellinger distance between signed measures; an application to product measures. Proc. Am. Math. Soc. **88**(4), 684–688 (1983). https://doi.org/10.2307/2045462

69. Stephanopoulos, N.O., McGhee, E.M.: Partisan gerrymandering and the efficiency gap. Univ. Chicago Law Rev. **82**(2), 831–900 (2015)

70. Szufa, S.: Optimal gerrymandering for simplified districts. Working Paper, Jagiellonian Center for Quantitative Research in Political Science, Kraków (2019)

71. Tideman, T.N., Plassmann, F.: Modeling the outcomes of vote-casting in actual elections. In: Felsenthal, D.S., Machover, M. (eds.) Electoral Systems. Paradoxes, Assumptions, and Procedures, pp. 217–251. Springer, Heidelberg (2012). https://doi.org/10.1007/978-3-642-20441-8_9

72. Tomz, M., Tucker, J.A., Wittenberg, J.: An easy and accurate regression model for multiparty electoral data. Polit. Anal. **10**(1), 66–83 (2002). https://doi.org/10.1093/pan/10.1.66

73. Tufte, E.R.: The relationship between seats and votes in two-party systems. Am. Polit. Sci. Rev. **67**(2), 540–554 (1973). https://doi.org/10.2307/1958782

74. Upton, G.: Blocks of voters and the cube 'law'. Br. J. Polit. Sci. **15**(3), 388–398 (1985). https://doi.org/10.1017/S0007123400004257

75. Wildgen, J.K., Engstrom, R.L.: Spatial distribution of partisan support and the seats/votes relationship. Legis. Stud. Q. **5**(3), 423–435 (1980). https://doi.org/10.2307/439554

76. Yamamoto, T.: A multinomial response model for varying choice sets, with application to partially contested multiparty elections. Working Paper, Massachusetts Institute of Technology, Cambridge, MA (2014)

77. Young, H.P.: Measuring the compactness of legislative districts. Legis. Stud. Q. **13**(1), 105–115 (1988). https://doi.org/10.2307/439947

Power in Networks: A PGI Analysis of Krackhardt's Kite Network

Manfred J. Holler[1] and Florian Rupp[2(✉)]

[1] Universitat Hamburg, Munich, Germany
[2] Center of Conflict Resolution (CCR), Untermeitingen, Germany
florian.rupp@ccr-munich.de

Abstract. This paper applies power index analysis to the well-known Krackhardt's kite social network by imposing a weighted voting game on the given network structure. It compares the results of this analysis, derived by applying the Public Good Index and the Public Value, with the outcome of employing the centrality concepts - degree centrality, closeness centrality, and betweenness centrality - that we find in Krackhardt (1990), and eigenvector centrality. The conclusion is that traditional centrality measures are rather a first approximation for evaluating the power in a network as they considerably abstract from decision making and thereby of possible coalitions and actions. Power index analysis takes care of decision making, however, in the rather abstract (a priori) form of the potential of forming coalitions.

Keywords: Network · Centrality · Public good index · Public value · Power indices · Weighted voting game

JEL Codes: C70 · C72 · D72 · D85 · L14 · Z13

1 Introduction and Preliminaries

Krackhardt (1990) introduced an example that challenges graph-theoretic centrality concepts of measuring the power of vertices in a kite-like network. In Krackhardt's kite network in Fig. 1, D has the highest vertex degree (degree centrality), H and I are essential for the connectivity of the network (betweenness centrality), and F and G have the average shortest path distance to the other vertices (closeness centrality).[1] The "kite structure" of Fig. 1 represents the smallest network Krackhardt has "found in which the centrality based on each of the three measures reveals different actors as the most central in the network." Later, several other authors introduced further examples for smallest networks with non-coinciding centralities, see Brandes and Hildenbrand (2014).

---

[1] Krackhardt refers to Freeman (1979) for definition and discussion of these concepts. E.g., degree centrality is defined as the number of links connected to the person. Closeness centrality is defined as the inverse of the average path distance between the actor and all others in the network. The definition of betweenness centrality needs a formal apparatus which will not be given here (see, e.g., Krackhardt 1990). See Sect. 4 of our paper.

© Springer-Verlag GmbH Germany, part of Springer Nature 2019
N. T. Nguyen et al. (Eds.): TCCI XXXIV, LNCS 11890, pp. 21–34, 2019.
https://doi.org/10.1007/978-3-662-60555-4_2

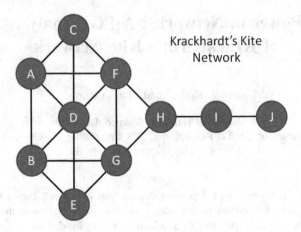

Fig. 1. Krackhardt's Kite Network

In this paper, we have chosen a more direct approach to measure the power of a vertex (or node) representing a decision maker (agent, player) in a network. It refers to the capacity of forming networks which is, of course, depending on the links between the vertices, i.e., the network structure. Given a specific network structure, we will consider coalitions that are minimal in as much as they contain only vertices, i.e., coalition members, which are critical to achieve the coalition's objective, e.g., building up a specific stock of resources necessary for financing a highway. An essential assumption of all standard power indices is that a critical decision maker – a "swing player" – has power. This holds for the indices of Shapley-Shubik, Penrose-Banzhaf-Coleman, Johnston and Deegan-Packel. We will apply the Public Good Index (PGI) as we focus networks that produce public goods.

The PGI represents the number of minimum winning coalitions (MWCs) which have a particular vertex i is an element – which is c_i – in the form of a ratio such that the shares of all vertices add up to one.[2] Thus, the PGI of i, given a particular network structure u and outcome rule d, is

$$h_i(u, d) = \frac{c_i}{c} \text{ with } \sum_{i=1}^{n} c_i = c \tag{1}$$

Here, c_i is a function of u and d. In the case of a collective decision problem d is the decision rule, e.g., a majority quorum.

Note that, because of public good assumption, there is "no splitting up of a cake" and no bargaining over shares as in Myerson (1977) and the contributions that build on it. Of course, there are networks that produce (private) goods that invite sharing, but in this study we focus on collective decision making over public goods. It is assumed that each winning coalition represents a particular public good. This is part of the story which motivated the application of the PGI.

---

[2] For a recent discussion of the PGI, see Holler (2019).

The PGI has been introduced in Holler (1982) and axiomatized in Holler and Packel (1983). However, as Holler and Li (1995) demonstrated, by looking at the shares only, relevant information can be lost. Therefore, we will also discuss the non-standardized numbers, i.e., the Public Value (PV): the PV of i is identical with the number of MWCs which have i as a member. Thus, the PV of i, given a particular net structure u and outcome rule d, is

$$p_i(u, d) = c_i \quad \text{such that} \quad \sum_{i=1}^{n} c_i = c \qquad (2)$$

Again, c_i is a function of the network structure u and the decision rule d. Holler and Li (1995) give an axiomatization of PV. In general, we will refer to c_i, the number of minimum winning coalitions when, in fact, we discuss its power interpretation PV. The PV measures the absolute power of a vertex while PGI measures the relative power.

In Sect. 2 we will discuss a voting game in which the players are linked in accordance to Krackhardt's kite network and two variations of it, applying PGI and PV. The effects on the distribution of changing the decision rule from simple majority to the 2/3 rule will be analyzed. In Sect. 3, we modify the kite structure and analyze the corresponding effects on the power measures, again applying simple majority and the 2/3 rule. Section 4 compares assumptions and results of the chosen power analysis to the centrality concepts chosen by Krackhardt (1990). Section 5 concludes the paper referring to some cognitive problems related to making decisions in network: e.g., do decision makers see and understand the links and how "deep" is this understanding, how many steps within a network do perception and comprehension cover?

2 Analyzing the Krackhardt's Kite Network

Let us consider Krackhardt's kite network in terms of a voting game v = (6; 1, 1, 1, 1, 1, 1, 1, 1, 1, 1) that has the absolute majority $d = 6$ as quorum,[3] and analyze it with a focus on the Public Good Index (PGI).[4] Vertices (nodes) are players in this game. A winning coalition is a set of *at least* six players. Given the network structure, players have to be *connected* (i.e., linked) to form a coalition. Thus a minimal winning coalition in the network game is a set of six players who are connected. There are 63 minimal winning coalitions which are satisfying the connectivity requirement implied by the network (see the Appendix for the listing).

We are in particular interested in environments more local and less global than those typically used in Graph Theory. By pure coincidence of the numbers, we are in this situation actually dealing with the small world properties of real networks (six degrees of separation).

---

[3] Think about a committee that decides about hiring a professor to the department. Nodes A to G represent the tight sub-network of incumbent professors, H, I and J are the representatives of the President of the University – also representing the bureaucratic personnel -, the assistants, and the students, respectively.

[4] Fragnelli (2013) analyzes a weighted voting game with network structure applying the Banzhaf (power) index.

To investigate the impact of the vertices I and J on the PGI-power of H, we consider the number of minimum winning coalitions – and calculate the corresponding PGIs – (a) in the complete graph Γ with vertices A to J, (b) those that are present in the sub-graph Γ_1 with vertices A to I (without vertex J), and (c) those that are present in the sub-graph Γ_2 with vertices A to H (without vertices I and J). These sub-graphs have the same edges as the complete graph otherwise. In Γ_1 there are 48 minimal winning coalitions present, and 25 in Γ_2. Table 1 gives the number c_i of minimum winning coalitions to which a vertex i belongs in these three graphs as well as the corresponding values of the Public Good Index (rounded to five decimals). Note that we did not reduce the quorum in the case of Γ_1 and Γ_2 in order to "isolate" effects of changes in the network structure.[5]

Table 1. Voting game v = (6; 1, 1, 1, 1, 1, 1, 1, 1, 1, 1) with network structures Γ, Γ_1, and Γ_2.

i	A	B	C	D	E	F	G	H	I	J
$c_i(\Gamma_2)$	19	19	18	18	18	20	20	18		
$PGI_i(\Gamma_2)$	0,12667	0,12667	0,12000	0,12000	0,12000	0,13333	0,13333	0,12000		
$c_i(\Gamma_1)$	32	32	28	34	28	35	35	41	23	
$PGI_i(\Gamma_1)$	0,11111	0,11111	0,09722	0,11806	0,09722	0,12153	0,12153	0,14236	0,07986	
$c_i(\Gamma)$	36	36	24	43	24	45	45	56	38	15
$PGI_i(\Gamma)$	0,09945	0,09945	0,06630	0,11878	0,06630	0,12431	0,12431	0,15470	0,10497	0,04144

In absolute terms, measured by the number of minimum winning coalitions c_i, F and G benefit if I joins H and even more so if J gets connected to I, i.e., their PV increases. In fact, the PV of all players increase if I gets connected to H. This is not surprising because the total number of elements in minimum winning coalitions, c, increases from 150 to 288. However, if J gets connected to I and the c-value increases further from 288 to 362, the PVs of C and E decreases. These two vertices become "peripherical" by the entry of J.

It seems obvious that H gains power if first I and then J gets connected. H's PV increases from (a relatively small) 18 to 41 and 56. The power gains and the prominent power position of H are also confirmed by the relative power captured by the PGI. H is the most powerful player in Krachhardt's kite network if I is connected with it, and even more so if J joins in. Correspondingly, all other vertices (with the exception of I and J) lose relative power in this process of extending the network to I and J as measured by the PGI. However, the relative power of F, G, and D recovers, at least to some extent, when J joins in addition to I – as this increases the chance to be in a minimum winning coalition for the three players; it seems that they benefit being close to H and thereby have a larger potential to connect with I and J than the nodes A, B, C, and E. The favorable PVs of F, G, and D are obvious from reading line $c_i(\Gamma)$ in Table 1.

---

[5] In general, parliaments do not change their majority rules if links between parties have increased or decreased, and, in the extreme, a party became unconnected to any other like going from Γ and Γ_1.

Krackhardt's kite network exhibits strong symmetries. In Γ_2 there are three groups of vertices were the elements of these groups have the same power: {A, B}, {C, D, H}, and {F, G}, i.e. A and B have, for instance, the same power concerning symmetric centrality measures. Being connected to H the group {F, G} has an advantage over {A, B}, because when introducing I and J, the vertices F and G can only participate in forming winning coalitions by the support of H, whereas the other vertices contribute equally. This broker position increases the power of H, or put differently: H has the power to exclude I and J from the political process. This reflects H's graph theoretic *betweenness centrality*.

In terms of political games F and G can only control their power by both excluding their participation in coalitions that contain I or {I, J}. Thus, H's connection to them and the increase in power for H they provide are not relevant, and the voting game reduces to a voting game on the sub-graph Γ_2.

Next, we study the voting game v = (7; 1, 1, 1, 1, 1, 1, 1, 1, 1, 1), again given Krackhardt's kite structure. The quorum of d = 7 reflects a 2/3 majority: this is a decision rule which is often relevant in changing a constitution or the voting rule itself. The number of minimal winning coalitions decreases to 39 (as listed in the Appendix) as there is of course a smaller potential for minimal winning coalitions as in the case of d = 6.

Table 2. Voting game v = (7; 1, 1, 1, 1, 1, 1, 1, 1, 1, 1) with network structures Γ, Γ_1, and Γ_2.

i	A	B	C	D	E	F	G	H	I	J
$c_i(\Gamma_2)$	7	7	7	7	7	7	7	7		
$PGI_i(\Gamma_2)$	0,12500	0,12500	0,12500	0,12500	0,12500	0,12500	0,12500	0,12500		
$c_i(\Gamma_1)$	16	16	14	15	14	17	17	19	12	
$PGI_i(\Gamma_1)$	0,11429	0,11429	0,10000	0,10714	0,10000	0,12143	0,12143	0,13571	0,08571	
$c_i(\Gamma)$	25	25	24	27	24	32	32	39	32	20
$PGI_i(\Gamma)$	0,08929	0,08929	0,08955	0,10075	0,08955	0,11940	0,11940	0,14552	0,11940	0,07463

At the first glance, the fact that all vertices in the Γ_2 network have the same PV $c_i(\Gamma_2) = 7$ is perhaps surprising. However, given Γ_2, the forming of minimal winning coalitions boils down to exclude one vertex out of eight. The eight vertices are well connected such that the exclusion of one node does not destroy the connectedness of any other.

When node I joins network Γ_2 to form Γ_1, the power of H increases and H becomes the most powerful node. This is not surprising as H is the gatekeeper for forming minimal winning coalitions that include I. What is however surprising is the differentiation among nodes A to G, although it can be concluded that F and G benefit from the closeness to the powerful H node while C, E, and D suffer from the fact that only 1 of them is needed to join if the other six nodes already form a proto-coalition. However, neither A nor B is needed to satisfy d = 7, if all the others agree to form a minimum winning coalition. Here it helps to look into the list of minimum winning coalitions and use the PV and PGI measures to get the results in Table 2.

This also holds for the unconstrained network Γ. Again, H is the most powerful node. Its neighbors F, G, and I are second. This supports the hypothesis that closeness to a "strong player" is beneficial to a "weaker player" – "strong" and "weak" defined by the PGI.

A comparison of Tables 1 and 2 reveals a rather substantial impact of the decision rule d on the distribution of power – most prominently perhaps in the equality of power for all nodes in Γ_2 in the case of d = 7, already mentioned. For the identical network structure, Table 1 shows some variation of power in the case of d = 6. Note also that the power of D is more "modest" in Table 2, still larger than the values of the neighboring C and E. The power of C and E is lowest in both settings, differentiated by d = 6 and d = 7, for networks Γ_1 and Γ. In the following, we will modify the network by erasing the direct links of D and C, and D and E and check the impact on the power distribution: whether the results discussed in this section still hold.

3   The D-Modified Kite Network

Algaba et al. (2018) define a pseudo-game u with player set $N \cup \Gamma$ where N is the set of vertices and Γ is the set of links. Links are players in this game.[6] Indeed, in the preceding we have seen that changes in the set of links have substantial consequences for the power distribution, if we measure power by PV and PGI. Of course, cutting the links of I, and I-J with node H is substantial for these vertices because they have no alternative: they are no longer connected and therefore are no longer candidates for a MWC within a network. Let us check the effect of a possibly less substantial modification in the network structure (again modifying the original pseudo-game by revising the set of links). We discuss a D-modified kite network, where the two links C-D and D-E of Krackhardt's kite network are erased (see Fig. 2.) Correspondingly we label the modified network structures by Γ°, Γ_{1°, and Γ_{2°. Again, to investigate the impact of the vertices I and J on the PGI-power of H, we consider the amount of minimally winning coalitions in the complete graph Γ° with vertices A to J, those that are present in the sub-graph Γ_{1° with vertices A to I (without vertex J), and those that are present in the sub-graph Γ_{2° with vertices A to H (without vertices I and J). These sub-graphs have the same edges as the complete graph otherwise.

In the D-modified kite network, F and G have the highest vertex degrees. Thus, F and G are, from a graph theoretic point of view, important due to degree centrality and closeness centrality. Though, as in the original kite network H turns out to be the PGI winner, given the voting game v = (6; 1, 1, 1, 1, 1, 1, 1, 1, 1, 1) on the network structures Γ°, Γ_{1°, and Γ_{2° (see Table 3). There are 55 minimal winning coalitions for this voting game if Γ° applies. Compared to Krackhardt's kite network the smaller degree of D has no impact on the number of minimum winning coalitions in Γ_2 as this sub-graph is already highly connected and six of the eight vertices therein are required for forming a minimally winning coalition which reduces the possible degrees of freedom for the formation and hence compensates for the fewer connections of D.

---

[6] For example, Aumann and Myerson (1988) identify links with players in the by now classical paper by Myerson (1977). However, links do not have preferences and do not gain payoffs.

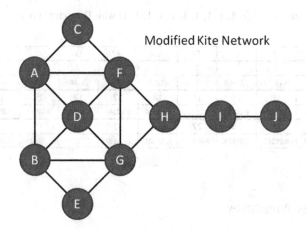

Fig. 2. D-modified Kite Network

Table 3. Voting game v = (6; 1, 1, 1, 1, 1, 1, 1, 1, 1, 1) with D-modified network structures Γ, Γ_1, and Γ_2.

i	A	B	C	D	E	F	G	H	I	J
$c_i(\Gamma_{2^\circ})$	19	19	18	18	18	20	20	18		
$PGI_i(\Gamma_{2^\circ})$	0,12667	0,12667	0,12000	0,12000	0,12000	0,13333	0,13333	0,12000		
$c_i(\Gamma_{1^\circ})$	30	30	24	28	24	32	32	35	17	
$PGI_i(\Gamma_{1^\circ})$	0,11905	0,11905	0,09524	0,11111	0,09524	0,12698	0,12698	0,13889	0,06746	
$c_i(\Gamma^\circ)$	34	34	27	35	27	41	41	48	30	13
$PGI_i(\Gamma^\circ)$	0,10303	0,10303	0,08181	0,10606	0,08181	0,12424	0,12424	0,14545	0,09091	0,03939

If we compare Tables 1 and 3 with respect to node I in the complete networks Γ and Γ°, then a loss of power seems obvious. For instance, in Γ, node I was "stronger" than nodes A or B, while after the D-modification I is weaker. Thus, changing the links of D has a rather substantial echo in the power of the "far-away" player I. On the other hand, node D, despite losing two links, does not suffer substantial power losses; it holds its number 4 position in the power ranking – if I and J enter. Again, there is quite an echo.

Should we expect similar effects for the voting game v = (7; 1, 1, 1, 1, 1, 1, 1, 1, 1, 1) with a quorum of 2/3? For the D-modified kite network we get 37 minimal winning coalitions for this game while for the Krackhardt's kite network we counted 39 minimum winning coalitions (see the Appendix for the listing).

Note, compared to Krackhardt's kite network the smaller degree of D has no impact on the number of minimum winning coalitions in Γ_{2° and Γ_{1°, i.e., a comparison of Tables 2 and 4 shows only rather small variations in the power values. A possible explanation for this result could be that for larger quorums the number of links of a player to not matter very much, because, in many configurations, it has to be included anyway to satisfy the quorum – and its neighbors will be included in the particular coalition irrespective of whether there is a direct link or a chain of connecting links.

Table 4. Voting game v = (7; 1, 1, 1, 1, 1, 1, 1, 1, 1, 1) with D-modified network structures Γ, Γ_1, and Γ_2.

i	A	B	C	D	E	F	G	H	I	J
$c_i(\Gamma_{2\cdot})$	7	7	7	7	7	7	7	7		
$PGI_i(\Gamma_{2\cdot})$	0,12500	0,12500	0,12500	0,12500	0,12500	0,12500	0,12500	0,12500		
$c_i(\Gamma_{1\cdot})$	16	16	14	15	14	17	17	19	12	
$PGI_i(\Gamma_{1\cdot})$	0,11429	0,11429	0,10000	0,10714	0,10000	0,12143	0,12143	0,13571	0,08571	
$c_i(\Gamma°)$	25	25	24	27	24	31	31	37	30	18
$PGI_i(\Gamma°)$	0,09191	0,09191	0,08824	0,09924	0,08824	0,11397	0,11397	0,13603	0,11029	0,06618

4 Centrality Measures

A plethora of centrality measures has been proposed, cf. Brandes and Hildenbrand (2014), Todeschini and Consonni (2009), and implemented in comprehensive software environments like R (see https://www.r-project.org/, especially the CINNA package). Due to their distinct nature four of these measures can be considered most promising in view of attributing power to members of a network: degree centrality, closeness centrality, betweenness centrality, and eigenvector centrality.[7]

Let us briefly recall the definitions of these centrality measures. The central assumption of *degree centrality* is that a vertex is important/powerful the more neighbors it has. In an undirected network, the degree of a vertex is the number of edges this vertex has. For instance, in Krackhardt's kite network vertex D has degree 6 and in particular thus the highest number of connections in the network and is the degree center therein, see Table 5.

Closeness centrality states that a vertex is important/ powerful if it has better access to information at other vertices or more direct influence on other vertices, cf. Freeman (1979). This means that the shortest network paths to other vertices are considered. Let us consider an undirected connected network with n vertices. Let $d_{i,j}$ be the length of the shortest path between vertex i and vertex j, then the mean distance l_i for vertex i reads $l_i = \frac{1}{n}\sum_{j=1}^{n} d_{i,j}$. By virtue of the definition, the mean distance is low for important vertices and high for unimportant ones. Therefore, its inverse is taken as closeness centrality C_i for vertex i reads

$$C_i = \frac{n}{\sum_{j=1}^{n} d_{i,j}}.$$

---

[7] For his analysis of the kite network Krackhardt applied degree centrality, closeness centrality, betweenness centrality only. His argument not to consider measures based on eigenvector centrality is that they "aimed more at the concept of asymmetric status hierarchy, or "being at the top", than they are at the idea of "being at the center", which is the idea behind the graph-theoretic measures used here" Krackhardt (1990: 351).

Table 5. Comparison of the values of four centrality measures in Krackhardt's kite network and its variations as discussed in Sect. 2 (computation with the software environment R).

	A	B	C	D	E	F	G	H	I	J
degree centrality										
Γ_2	4	4	3	6	3	5	5	2		
Γ_1	4	4	3	6	3	5	5	3	1	
Γ	4	4	3	6	3	5	5	3	2	1
closeness centrality										
Γ_2	0.10000000	0.10000000	0.09090909	0.12500000	0.09090909	0.11111111	0.11111111	0.08333333		
Γ_1	0.07692308	0.07692308	0.07142857	0.09090909	0.07142857	0.09090909	0.09090909	0.07692308	0.05000000	
Γ	0.05882353	0.05882353	0.05555556	0.06666667	0.05555556	0.07142857	0.07142857	0.06666667	0.04761905	0.03448276
betweennes centrality										
Γ_2	0.8333333	0.8333333	0.0000000	3.6666667	0.0000000	3.3333333	3.3333333	0.0000000		
Γ_1	0.8333333	0.8333333	0.0000000	3.6666667	0.0000000	5.8333333	5.8333333	7.0000000	0.0000000	
Γ	0.8333333	0.8333333	0.0000000	3.6666667	0.0000000	8.3333333	8.3333333	14.0000000	8.0000000	0.0000000
eigenvector centrality										
Γ_2	0.7326933	0.7326933	0.5943841	1.0000000	0.5943841	0.8216742	0.8216742	0.3823961		
Γ_1	0.73223962	0.73223962	0.59423476	1.00000000	0.59423476	0.82647438	0.82647438	0.40576524	0.09423476	
Γ	0.73221232	0.73221232	0.59422577	1.00000000	0.59422577	0.82676381	0.82676381	0.40717690	0.09994054	0.02320742

Table 6. Comparison of the values of four centrality measures in the D-modified kite network and its variations as discussed in Sect. 3 (computation with the software environment R).

	A	B	C	D	E	F	G	H	I	J
degree centrality										
Γ_{2^*}	4	4	2	4	2	5	5	2		
Γ_{1^*}	4	4	2	4	2	5	5	3	1	
Γ°	4	4	2	4	2	5	5	3	2	1
closeness centrality										
Γ_{2^*}	0.10000000	0.10000000	0.07692308	0.10000000	0.07692308	0.11111111	0.11111111	0.08333333		
Γ_{1^*}	0.07692308	0.07692308	0.06250000	0.07692308	0.06250000	0.09090909	0.09090909	0.07692308	0.05000000	
Γ°	0.05882353	0.05882353	0.05000000	0.05882353	0.05000000	0.07142857	0.07142857	0.06666667	0.04761905	0.03448276
betweennes centrality										
Γ_{2^*}	2.3333333	2.3333333	0.0000000	0.6666667	0.0000000	4.8333333	4.8333333	0.0000000		
Γ_{1^*}	2.3333333	2.3333333	0.0000000	0.6666667	0.0000000	7.3333333	7.3333333	7.0000000	0.0000000	
Γ°	2.3333333	2.3333333	0.0000000	0.6666667	0.0000000	9.8333333	9.8333333	14.0000000	8.0000000	0.0000000
eigenvector centrality										
Γ_{2^*}	0.8644546	0.8644546	0.4853771	0.9707542	0.4853771	1.0000000	1.0000000	0.5206639		
Γ_{1^*}	0.8557725	0.8557725	0.4812972	0.9625944	0.4812972	1.00000000	1.00000000	0.5561084	0.1442275	
Γ°	0.85510976	0.85510976	0.48098559	0.96197117	0.48098559	1.00000000	1.00000000	0.55882611	0.15533232	0.04027395

Betweenness centrality considers the power of a vertex by means of its control over information passing between other vertices, cf. Brandes (2001), Freeman 1979. In an undirected connected network, this means that the betweenness centrality measure b_i focuses on the extent to which a vertex i lies on paths between other vertices s and t

$$b_i = \sum_{s,t} \frac{n_{s,t}^i}{n_{s,t}},$$

where $n_{s,t}$ is the total number of (directed) paths from s to t, and $n_{s,t}^i$ denotes the number of (directed) paths from s to t that pass through vertex i.

Eigenvector centrality assumes a vertex to be important/powerful if it is connected to other important/powerful vertices, cf. Bonacich (1987). Mathematically, this means to solve an eigenvalue problem

$$\lambda x = A x,$$

where A is the adjacency matrix of the network. Assumed that A is a real-valued matrix with non-negative entries and the property that a power A^k, $K \geq 1$, has positive entries only, then the Perron-Frobenius Theorem, guarantees that a positive eigenvalue λ with algebraic multiplicity one exists such that its absolute value is larger than the absolute value of any other eigenvalue of A. Especially, there is an eigenvector x corresponding to λ having positive entries only. The entries of this eigenvector give a natural ordering for the importance/ power of a vertex.

It was Krackhardt's intention to demonstrate that the three centrality concepts – degree centrality, closeness centrality, and betweenness centrality – pick three different vertices as "winners" for his kite network; in fact, he proposed the particular kite structure in order to show this result (Krackhardt 1990: 351). Here the eigenvector centrality supports degree centrality, however, it seems obvious that this is not always the case. By and large, the power index analysis supports betweenness centrality: both concepts favor vertex H which is not surprising as, in many configurations of the network, it can function as a sort of gatekeeper exerting corresponding power.

5   Centrality Versus Public Good Index

The application of these four centrality measures on Krackhardt's kite network, the D-modified network and their variations are given in Tables 5 and 6. Compared to the values of the PGI from Sects. 2 and 3, we see that, not surprisingly, traditional graph-theoretic centrality measures fail in recovering voting power in networks (which is here expressed in terms of the PGI). What can be recognized is that the centrality winner is in general rather stable with respect to adding I or I-J. Moreover, the graph-theoretic centrality concept closest to the idea of the PGI is betweenness centrality. Although, the voting game quorum means that not all paths are considered, but only those with the fixated length defined by the quorum, and that the respective vertex is allowed to be the *initial or terminal vertex of the path as well.*

By considering power in networks from the perspective of voting games, we introduced a third point of view and paradigm of centrality. The first class of centrality concepts developed were those related to graph theory, cf. Krackhardt (1990), and that aimed to choose one or a rather small group of central vertices such that the fundamental properties would considerably change without this vertex, cf. Barabasi (2016);

Newmann (2010). In particular, in view of early military applications of graph theory, a vertex that is connected to a lot of others presents itself as a profitable target for bombers. Further, a second class of centrality concepts aims to describe status hierarchies in social networks, cf. Hubbell (1965); Bonacich (1987), and Salancik and Pfeffer (1986). Examples include the discussed eigenvalue centrality or a centrality concept that interprets power in the sense of executing power with respect to unimportant vertices as discussed by Bozzo and Franceschet (2016). Depending on the actual situation these centrality concepts coincide with the former graph theoretic ones or are more tailored in the sense of including specifics of social networks. Our third class of centrality concepts relates power to coalition formation in voting games. As already stated, computationally the PGI on networks is connatural to a quorum ramified betweenness centrality, where all paths of a fixed quorum-length are taken into account that include a specific vertex, and where this vertex is allowed to be the initial or terminal vertex of the paths.

Above we made attempts to compare the results of applying the PGI, on the one hand, and centrality concepts, on the other. Of course, it would be interesting to have a more fundamental comparison of the power index approach and the centrality concepts. However, we have to accept that the two approaches are very different – e.g., the centrality concepts try to express power without reference to a decision problem while, in a network environment, power index analysis refers to pseudo-games, as defined above, with player sets $N \cup \Gamma$ where N is the set of vertices and Γ is the set of links. Those who control the links have power. In the above analysis we assumed that links between vertices i and j are controlled by i and j. However, quite often the link between i is controlled by k, a third agent many – then k has a potential to exert power. That is why we pay our tolls to telephone companies. Degree centrality, closeness centrality, and betweenness centrality are not designed to take care of this issue – a shortcoming for these concepts if applied to express power in networks. In fact, these concepts look like a first approximation when it comes to power. How can we measure the power of a vertex in a network if the specification and dedication of the network are unknown (or not defined) – e.g., if we do not know whether it is an information network, a distribution network or an ideological network underlying a voting institution. In "bargaining situations it is advantageous to be connected to those who have few options; power comes from those being connected to those who are powerless" (Bonacich 1987: 1171) while an information network is, in general, advantageous if we are connected to many "who know." Above we have chosen a voting model to specify the network; links defined possible paths of coalitional decision making[8] – power relations. Alternatively, we could have engrafted a bargaining model of the Myerson type (see Myerson 1977; Aumann and Myerson 1988) onto Krackhardt's kite structure.

---

[8] Krackhardt (1990) did not consider decision making. He focused on the cognitive problem of what network members know about the network and about other members of a network. Degree centrality, closeness centrality, and betweenness centrality might be reasonable instrument to evaluate one's network position and the positions of the others – in fact, to recognize a network. "The central point" in his paper, however, is: "Cognitive accuracy of the informal network is, in and of itself, a base of power" (Krackhardt 1990: 343). The power index analysis dealt primarily with the formal structure, however, the links between the various nodes could be highly informal.

Krackhardt (1990) contains a real-world example of a firm of 36 employees, including the three top managers who own the company. Qualifications of competence and charisma, revealed by means of questionnaires, are added to the structure of interaction at work to get from centrality to (reputational) power. This is an alternative to assigning a game – which makes perfect sense if the agents are not expected to behave strategically and, e.g., coalition formation does not matter.

Appendix

1. Set of minimal winning coalitions of the voting game v = (6; 1, 1, 1, 1, 1, 1, 1, 1, 1, 1) given Krackhardt's kite network:

{J, I, H, F, C, A}	{J, I, H, G, D, C}	{I, H, F, D, G, B}	{H, G, F, C, D, A}	{H, F, C, A, B, D}
{J, I, H, F, C, D}	{J, I, H, G, F, A}	{I, H, F, D, G, E}	{H, G, F, C, D, B}	{H, F, C, A, B, E}
{J, I, H, F, C, G}	{I, H, F, C, A, B}	{I, H, G, E, B, A}	{H, G, F, C, E, A}	{H, F, C, D, B, E}
{J, I, H, F, D, A}	{I, H, F, C, A, D}	{I, H, G, E, B, D}	{H, G, F, C, E, B}	{H, F, A, D, B, E}
{J, I, H, F, D, B}	{I, H, F, C, A, G}	{I, H, G, E, B, F}	{H, G, F, C, A, B}	{E, B, G, A, C, F}
{J, I, H, F, D, G}	{I, H, F, C, D, E}	{I, H, G, E, D, C}	{H, G, F, E, A, B}	{E, B, G, A; D, F}
{J, I, H, F, D, E}	{I, H, F, C, D, B}	{I, H, G, B, A, D}	{H, G, F, E, B, D}	{E, B, G, F, D, C}
{J, I, H, F, G, B}	{I, H, F, A, D, G}	{I, H, G, B, D, C}	{H, G, F, E, D, A}	{E, B, G, A, D, C}
{J, I, H, F, G, E}	{I, H, F, A, D, E}	{I, H, G, B, A, C}	{H, G, F, D, A, B}	{E, B, A, C, D, F}
{J, I, H, G, E, B}	{I, H, F, A, B, D}	{I, H, G, D, A, C}	{H, G, E, A, B, D}	{E, G, A, C, D, F}
{J, I, H, G, E, D}	{I, H, F, A, B, G}	{I, H, G, D, A, E}	{H, G, E, A, B, C}	{C, A, F, D, B, G}
{J, I, H, G, D, B}	{I, H, F, A, B, E}	{I, H, G, D, F, C}	{H, G, E, D, A, C}	
{J, I, H, G, D, A}	{I, H, F, D, B, E}	{H, G, F, C, D, E}	{H, G, B, D, A, C}	

2. Set of minimal winning coalitions of the voting game v = (7; 1, 1, 1, 1, 1, 1, 1, 1, 1, 1), representing a 2/3 quorum, given Krackhardt's kite network:

{A, B, C, D, E, F, G}	{I, H, F, C, A, B, D}	{I, H, F, G, D, C, E}	{J, I, H, G, E, B, A}	{J, I, H, F, G, A, E}
{A, B, C, D, E, F, H}	{I, H, G, E, B, A, C}	{I, H, F, G, D, E, A}	{J, I, H, G, E, B, D}	{J, I, H, F, G, B, C}
{A, B, C, D, E, G, H}	{I, H, G, E, B, A, D}	{I, H, F, G, D, E, B}	{J, I, H, G, E, D, A}	{J, I, H, F, G, A, D}
{A, B, C, D, F, G, H}	{I, H, F, G, A, B, C}	{J, I, H, F, C, A, B}	{J, I, H, G, D, A, C}	{J, I, H, F, G, D, C}
{A, B, C, E, F, G, H}	{I, H, F, G, A, B, D}	{J, I, H, F; C, A, D}	{J, I, H, G, D, C, E}	{J, I, H, F, G, D, E}
{A, C, D, E, F, G, H}	{I, H, F, G, A, B, E}	{J, I, H, F, C, D, B}	{J, I, H, F, G, C, A}	{J, I, H, F, G, D, B}
{B, C, D, E, F, G, H}	{I, H, F, G, D, C, A}	{J, I, H, F, C, D, E}	{J, I, H, F, G, A, B}	{J, I, H, F, G, C, E}
{I, H, F, C, A, B, E}	{I, H, F, G, D, C, B}	{J, I; H, F, D, B, E}	{J, I, H, F, G, B, E}	

3. Set of minimal winning coalitions of the voting game v = (6; 1, 1, 1, 1, 1, 1, 1, 1, 1, 1), given the D-modified Krackhardt's kite network:

{J, I, H, F, C, A}	{J, I, H, G, D, A}	{I, H, F, D, G, E}	{H, G, F, C, E, A}	{H, F, C, A, B, D}
{J, I, H, F, C, D}	{J, I, H, G, F, A}	{I, H, G, E, B, A}	{H, G, F, C, E, B}	{H, F, C, A, B, E}
{J, I, H, F, C, G}	{I, H, F, C, A, B}	{I, H, G, E, B, D}	{H, G, F, C, A, B}	{H, F, C, D, B, E}
{J, I, H, F, D, A}	{I, H, F, C, A, D}	{I, H, G, E, B, F}	{H, G, F, E, A, B}	{H, F, A, D, B, E}
{J, I, H, F, D, B}	{I, H, F, C, A, G}	{I, H, G, B, A, D}	{H, G, F, E, B, D}	{E, B, G, A, C, F}
{J, I, H, F, D, G}	{I, H, F, A, D, G}	{I, H, G, B, A, C}	{H, G, F, E, D, A}	{E, B, G, A, D, F}
{J, I, H, F, G, B}	{I, H, F, A, B, D}	{I, H, G, D, A, C}	{H, G, F, D, A, B}	{E, B, G, F, D, C}
{J, I, H, F, G, E}	{I, H, F, A, B, G}	{I, H, G, D, F, C}	{H, G, E, A, B, D}	{E, B, G, A, D, C}
{J, I, H, G, E, B}	{I, H, F, A, B, E}	{H, G, F, C, D, E}	{H, G, E, A, B, C}	{E, B, A, C, D, F}
{J, I, H, G, E, D}	{I, H, F, D, B, E}	{H, G, F, C, D, A}	{H, G, E, D, A, C}	{E, G, A, C, D, F}
{J, I, H, G, D, B}	{I, H, F, D, G, B}	{H, G, F, C, D, B}	{H, G, B, D, A, C}	{C, A, F, D, B, G}

4. Set of minimal winning coalitions of the voting game v = (7; 1, 1, 1, 1, 1, 1, 1, 1, 1, 1), representing a 2/3 quorum, given the D-modifiedKrackhardt's kite network:

{A, B, C, D, E, F, G}	{I, H, F, C, A, B, D}	{I, H, F, G, D, C, E}	{J, I, H, G, E, B, A}	{J, I, H, F, G, A, E}
{A, B, C, D, E, F, H}	{I, H, G, E, B, A, C}	{I, H, F, G, D, E, A}	{J, I, H, G, E, B, D}	{J, I, H, F, G, B, C}
{A, B, C, D, E, G, H}	{I, H, G, E, B, A, D}	{I, H, F, G, D, E, B}	{J, I, H, G, E, D, A}	{J, I, H, F, G, A, D}
{A, B, C, D, F, G, H}	{I, H, F, G, A, B, C}	{J, I, H, F, C, A, B}	{J, I, H, G, D, A, C}	{J, I, H, F, G, D, C}
{A, B, C, E, F, G, H}	{I, H, F, G, A, B, D}	{J, I, H, F; C, A, D}	{J, I, H, F, G, C, A}	{J, I, H, F, G, D, E}
{A, C, D, E, F, G, H}	{I, H, F, G, A, B, E}	{J, I, H, F, C, D, B}	{J, I, H, F, G, A, B}	{J, I, H, F, G, D, B}
{B, C, D, E, F, G, H}	{I, H, F, G, D, C, A}	{J, I; H, F, D, B, E}	{J, I, H, F, G, B, E}	{J, I, H, F, G, C, E}
{I, H, F, C, A, B, E}	{I, H, F, G, D, C, B}			

References

Algaba, E., López, S., Owen, G., Saboyá, M.: A game-theoretic approach to networks (2018). manuscript

Aumann, R., Myerson, R.B.: Endogenous formation of links between players and coalitions: an application of the shapley value. In: Roth, A. (ed.) the shapley value, pp. 175–191. Cambridge University Press, Cambridge (1988)

Barabasi, A.-L.: Network Science. Cambridge University Press, Cambridge (2016)

Bonacich, P.: Power and centrality: a family of measures. Am. J. Sociol. 92, 1170–1182 (1987)

Bozzo, E., Franceschet, M.: A theory on power in networks. Commun. ACM 59, 75–83 (2016)

Brandes, U.: A faster algorithm for betweenness centrality. J. Math. Sociol. 25, 163–177 (2001)

Brandes, U., Hildenbrand, J.: Smallest graphs with distinct singleton centers. Netw. Sci. 2, 416–418 (2014)

Fragnelli, V.: A note on communication structures. In: Holler, M.J., Nurmi, H. (eds.) Power, Voting, and Voting Power: 30 Years After, pp. 467–473. Springer, Heidelberg (2013). https://doi.org/10.1007/978-3-642-35929-3_24

Freeman, L.C.: Centrality in social networks: conceptual clarification. Soc. Netw. 1, 215–239 (1979)

Holler, M.J.: Forming coalitions and measuring voting power. Polit. Stud. **30**, 262–271 (1982)

Holler, M.J.: The story of the poor public good index. Transaction on Computational Collective Intelligence (forthcoming) (2019)

Holler, M.J., Packel, E.W.: Power, luck, and the right index. J. Econ. (Zeitschrift für Nationalökonomie) **43**, 21–29 (1983)

Holler, M.J., Li, X.: From public good index to public value: an axiomatic approach and generalization. Control Cybern. **24**, 257–270 (1995)

Hubbell, C.H.: An input-output approach to clique identification. Sociometry **28**, 377–399 (1965)

Krackhardt, D.: Assessing the political landscape: structure, cognition, and power in organizations. Adm. Sci. Q. **35**, 342–369 (1990)

Myerson, R.B.: Graphs and cooperation in games. Math. Oper. Res. **2**, 225–229 (1977)

Newmann, M.E.J.: Networks: An Introduction. Oxford University Press, Oxford (2010)

Salancik, G.R., Pfeffer, J.: Who gets power and how they hold onto it: a strategic-contingency model of power. Organ. Dyn. **5**, 3–21 (1986)

Todeschini, R., Consonni, V.: Molecular Descriptors for Chemoinformatics, Volumes I & II. Wiley-VCH, Weinheim (2009)

Orders of Criticality in Graph Connection Games

Marco Dall'Aglio[1], Vito Fragnelli[2(✉)], and Stefano Moretti[3]

[1] Department of Economics and Finance, LUISS University,
Viale Romania 32, 00197 Rome, Italy
mdallaglio@luiss.it
[2] Department of Sciences and Innovative Technologies (DISIT),
University of Eastern Piedmont, Viale T. Michel 11, 15121 Alessandria, Italy
vito.fragnelli@uniupo.it
[3] Université Paris-Dauphine, PSL Research University, CNRS, LAMSADE,
Place du Maréchal de Lattre de Tassigny, 75775 Paris Cedex 16, France
stefano.moretti@dauphine.fr

Abstract. The order of criticality of a player in a simple game and two indices inspired by the reasoning à la Shapley and à la Banzhaf were introduced in two previous papers [3] and [4], respectively, mainly having in mind voting situations. Here, we devote our attention to graph connection games, and to the computation of the order of criticality of a player. The indices introduced in [4] may be used as centrality measures of the edges in preserving the connection of a graph.

Keywords: Order of criticality · Shapley value · Banzhaf value · Graph connection games

1 Introduction

How crucial is a player in making a coalition win? This characteristic is traditionally measured by the notion of critical player, whose intervention changes the sorts of a given coalition. In turn, this idea leads to the building of power measures such as the Shapley-Shubik, or the Bahnzaf power indices. These indicators do not take into account the possibility for an agent to interact with other in the making or in the dismantling of a winning coalition. For this reason, the order of criticality of a player in a simple game was defined in [3], then two indices inspired by the reasoning à la Shapley and à la Banzhaf, i.e. accounting for the ordering in which the players enter a coalition or simply accounting for the coalition were introduced in [4].

In those papers, we devoted our interest mainly to voting situations; here, we want to extend the previous results to a different class of simple games, the so-called graph connection games.

The rest of the paper is organized as follows: Sect. 2 contains the basic notions and notations we use throughout the paper; in Sect. 3 graph connection games

© Springer-Verlag GmbH Germany, part of Springer Nature 2019
N. T. Nguyen et al. (Eds.): TCCI XXXIV, LNCS 11890, pp. 35–46, 2019.
https://doi.org/10.1007/978-3-662-60555-4_3

are presented and some characteristics of the order of criticality are given; Sect. 4 is devoted to the computation of the order of criticality; in Sect. 5 we examine some classes of graphs for which the computation of the order of criticality is simple; in Sect. 6 we recall the collective indices introduced in [4] and apply them to the graph connection games; Sect. 7 concludes.

2   Preliminaries

Let us start with some preliminaries. An *undirected graph* is a pair $G = (V, A)$ where $V = \{v_1, v_2, ..., v_m\}$ is the finite set of *vertices* and $A = \{a_1, a_2, ..., a_n\}$ is the set of *edges*, i.e. pairs of vertices called *extremes* of the edge. For sake of simplicity, we consider only *simple* graphs, i.e. graphs in which there exists at most one edge among each pair of vertices. Given a graph $G = (V, A)$, a *path* among two vertices v_i and v_j is a sequence of distinct edges such that each edge has one extreme in common with the preceding one in the sequence and the other extreme in common with the following one in the sequence; v_i is the extreme of the first edge of the sequence not in common with the second edge and v_j is the extreme of the last edge of the sequence not in common with the second last edge. Given a graph $G = (V, A)$, a *cycle* is path in which the two extreme vertices coincide and there is no vertex that is in common to more than two edges; in other words, the path crosses each vertex at most once.

A graph is *connected* if there exists a path among each pair of vertices. Given a graph $G = (V, A)$, a *subgraph* is another graph $G' = (V', A')$ where $V' \subseteq V$ and $A' \subseteq A$. If the set V' is not specified, but only $A' \subseteq A$ is given, in the following we will refer to the subgraph $G_{A'} = (V, A')$, i.e. a graph with the same set of vertices and the subset of edges A'. An *induced subgraph* of $G = (V, A)$ is a graph $G^* = (V^*, A^*)$ formed by a subset of the vertices $V^* \subseteq V$ and such that A^* contains all the edges in A connecting pairs of vertices in V^*. A subgraph of G is a *spanning tree* if it is connected and does not contain cycles.

A *cooperative game with transferable utility* (TU-game) is a pair (N, v), where $N = \{1, 2, ..., n\}$ denotes the *finite set of players* and $v : 2^N \to \mathbb{R}$ is the *characteristic function*, with $v(\varnothing) = 0$. $v(S)$ is the worth of coalition $S \subseteq N$, i.e. what players in S may obtain standing alone.

A TU-game (N, v) is *simple* when $v : 2^N \to \{0, 1\}$, with $S \subseteq T \Rightarrow v(S) \leq v(T)$[1] and $v(N) = 1$. If $v(S) = 0$ then S is a *losing* coalition, while if $v(S) = 1$ then S is a *winning* coalition. Given a winning coalition S, if $S \setminus \{i\}$ is losing then $i \in N$ is a *critical player* for S. When a coalition S contains at least one critical player for it, S is a *quasi-minimal winning coalition*; when all the players of S are critical, it is a *minimal winning coalition*. A simple game may be defined also assigning the set of winning coalitions or the set of minimal winning coalitions.

---

[1] This property is called *monotonicity*.

Given a simple game (N, v), we recall the definition of criticality given in [3]:

Definition 1. *Let $k \geq 0$ be an integer, let $M \subseteq N$, with $|M| \geq k + 1$, be a winning coalition. We say that player i is* critical *of the order $k + 1$ for coalition M, via coalition K, with $|K| = k$ iff $K \subseteq M \setminus \{i\}$ is a set of minimal cardinality such that*

$$v(M \setminus K) - v(M \setminus (K \cup \{i\})) = 1 \tag{1}$$

The meaning of the definition is that K is a coalition of minimal cardinality such that $M \setminus K$ is still a winning coalition, while $M \setminus (K \cup \{i\})$ becomes a losing one[2].

In the above definition, coalition K is often omitted and we simply say that i is critical of a certain order for coalition $M \subseteq N$. Moreover, if no confusion arises, when $M = N$, we say that i is critical of a certain order in game (N, v).

Notice also that, when $k = 0$, $K = \varnothing$, thus, a player is critical of order 1 if and only if it is critical in the usual sense.

The following results hold:

(i) Let $M \subseteq N$ be a winning coalition, then the players in M may be partitioned according to their order of criticality, possibly including the subset of those players that are never critical, if any (Corollary 1 in [3]).

(ii) Let $i \in M$ be a player critical of the order $k + 1, k \geq 1$ for coalition M, via coalition $K \subset M$; if a player $j \in K$ leaves the coalition, then i is a player critical of the order k for coalition $M \setminus \{j\}$, via coalition $K \setminus \{j\}$.

Example 1. *Consider a simple game (N, v) with $N = \{1, 2, 3, 4, 5, 6\}$ and minimal winning coalitions $\{1, 2\}$, $\{1, 3, 4\}$, $\{1, 3, 5\}$. Player 1 is critical of the first order; players 2 and 3 are critical of the second order (2 is critical via $\{3\}$ and 3 is critical via $\{2\}$); players 4 and 5 are critical of the third order (4 is critical via $\{2, 5\}$ and 5 is critical via $\{2, 4\}$); player 6 is never critical.*

We may get further information from the dual game. We recall that given a game (N, v), we may define its dual game (N, v^*) that has the same set of players N and whose characteristic function is defined as $v^*(S) = v(N) - v(N \setminus S)$ for each $S \subseteq N$.

It is easy to check that if player $i \in N$ is critical of order k via coalition $K \subseteq N \setminus \{i\}$, then $v^*(K) = 0$ and $v^*(K \cup \{i\}) = 1$, i.e. player i is critical in the usual sense for coalition K.

3   Graph Connection Games

Given a connected simple graph $G = (V, A)$ we define the Graph Connection (GC) game associated to G as the simple TU-game (N, v) where $N = \{1, 2, ..., n\}$

---

[2] In other terms, $v(M \setminus T) = 0$ or $v(M \setminus (T \cup \{i\})) = 1$ for any $T \subset M \setminus \{i\}$ with $|T| < k$.

is the finite set of players and coincides with the set of the edges, i.e. edge $a_i \in A$ is player $i \in N$, and $v : 2^N \rightarrow \{0, 1\}$ is the characteristic function, defined as $v(S) = 1$ if the subgraph that contains all the vertices in V but only the edges owned by the agents in $S \subseteq N$ is connected and $v(S) = 0$ otherwise (see [1]).

The notion of criticality order for these games can be used to measure the strength of a part of a connected system. Suppose that the edges are bridges of a city under attack that has to remain connected in order to survive. The customary notions may help us distinguish between critical bridges, whose distruction would doom the city, and the non-critical ones, that can be the target of a single attack without harming the general system. Introducing higher levels of criticality enables us to measure the importance of each single bridge under coordinate attacks that involve the simultaneaous distruction of several bridges.

Differently from the simple games arising from weighted majority games, that were studied in [4], for Graph Connection Games we may specify some characteristics of the order of criticality:

(i) The minimal winning coalitions correspond to the spanning trees of the given graph, so they have the same cardinality that is $m - 1$, where m is the number of vertices.

(ii) There do not exist players that are never critical and the maximum order of criticality of a player is $n - m + 2$, i.e. 1 plus the number of edges that have to be eliminated in order to obtain a spanning tree.

Let us summarize the previous concepts with the following example.

Example 2. *Consider the following graph, where the number at each edge indicates the player.*

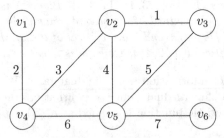

The minimal winning coalitions are $\{1, 2, 3, 4, 7\}$, $\{1, 2, 3, 5, 7\}$, $\{1, 2, 3, 6, 7\}$, $\{1, 2, 4, 6, 7\}$, $\{1, 2, 5, 6, 7\}$, $\{2, 3, 4, 5, 7\}$, $\{2, 3, 5, 6, 7\}$, $\{2, 4, 5, 6, 7\}$ *that correspond to the spanning trees obtained by eliminating one edge for each of the two cycles* $\{v_2, v_3\}$, $\{v_3, v_5\}$, $\{v_5, v_2\}$ *and* $\{v_2, v_4\}$, $\{v_4, v_5\}$, $\{v_5, v_2\}$.
Players $2, 7$ *that belong to no cycle are critical of order 1 as they are bridges, players* $1, 3, 5, 6$ *that belong to a unique cycle are critical of order 2 (1 is critical via* $\{5\}$ *and 5 is critical via* $\{1\}$, *3 is critical via* $\{6\}$ *and 6 is critical via* $\{3\}$*),
player 4 that belongs to both cycles is critical of order 3 (4 is critical via* $\{1, 3\}$, $\{1, 6\}$, $\{3, 5\}$ *or* $\{5, 6\}$*).
Note also that the order of criticality of player 4 is equal to* $n - m + 2 = 7 - 6 + 2 = 3$.

Example 3. *Consider the following graph.*

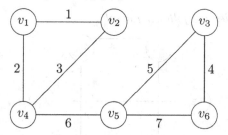

Player 6 that belongs to no cycle is critical of order 1, while the other players that belong to a unique cycle are critical of order 2.
Note that no player has the maximal order of criticality $n - m + 2 = 7 - 6 + 2 = 3$.

Starting from a graph connection game, a possible interpretation of the dual game is that a coalition $S \subseteq N$ is winning in the dual game if and only if $v(N \setminus S) = 0$, i.e. if the agents of S are able to disconnect the graph, or in other words if they are able to "sabotage" the connection of the graph.

4   The Computation

In this section, we consider the problem of the computation of the order of criticality of a player in graph connection games.

To compute the order of criticality associated to an edge $\{v_i, v_j\}$ in a winning coalition S, we need to find the minimal number of other edges in S that, together with $\{v_i, v_j\}$, disconnect, i.e. cut, the graph, leaving the two vertices v_i and v_j on different sides of the cut. The problem can be therefore formulated as a classical minimum cut problem in graph theory, by means of the following simple steps:

1. Remove $\{v_i, v_j\}$ from S;
2. Transform each undirected edge as two directed edges in opposite direction with the same unitary capacity;
3. Compute the capacity k of a minimum cut from v_i to v_j (or viceversa), by means of the Ford-Fulkerson algorithm (see [5]);
4. The order of criticality is given by $k + 1$.

We briefly outline the Ford-Fulkerson algorithm:

1. Let $k = 0$;
2. The cycle:
   (a) Find an augmenting path from v_i to v_j, if there is none exit, returning k;
   (b) Compute the maximum flow f passing through this path;
   (c) Now transform the graph into its residual by modifying the capacities of the edges composing the augmenting path, and those in the opposite direction as follows:
      • The capacity of each edge in the path is reduced by f;
      • The corresponding edge in the opposite direction is augmented by f;
   (d) Let $k := k + 1$ and repeat.

We give two examples of its functioning.

Example 4. *Consider the following graph.*

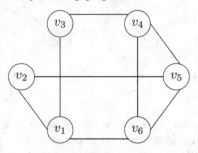

We want to compute the order of criticality of the edge $\{v_1, v_6\}$, so we remove it, transform the other edges as two directed arcs in opposite directions with unitary capacity (we do not indicate the capacity of arcs when it is 1) and set $k = 0$.

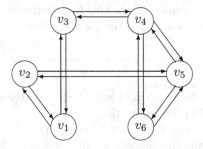

First augmenting path

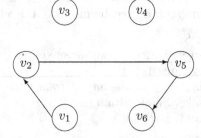

Residual graph (we indicate only the capacity of arcs different from 1); $k = 1$

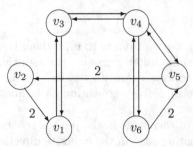

Second augmenting path

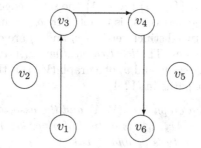

Residual graph; k = 2

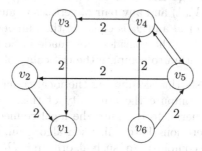

There do not exist other augmenting paths from v_1 to v_6, so the order of criticality of the edge $\{v_1, v_6\}$ is $k + 1 = 3$.

In the following example, we show what happens when the order of criticality of an edge is 1.

Example 5. *Consider the following graph.*

We want to compute the order of criticality of the edge $\{v_2, v_3\}$, so we remove it, transform the other edges and set $k = 0$.

There do not exist augmenting paths from v_2 to v_3, so the order of criticality of the edge $\{v_2, v_3\}$ is $0 + 1 = 1$.

5   Local Connectivity

In this section we investigate the relation between the order of criticality of a player in a graph $G = (V, A)$ connection games and the notion of *local connectivity* for edges in G. We first need to recall some further definitions from graph theory. For any vertex $v_i \in V$, we denote by $N_{v_i}^G$ the set of *neighbors* of v_i in

(V, A), i.e. the set of all vertices adjacent with i in (V, A), by $d^G(v_i) = |N_{v_i}^G|$ the *degree* of v_i and by $A_i^G = \{\{v_i, v_j\} : v_j \in A\}$ the set of edges incident to v_i. The minimum degree over the vertices V is denoted by $\delta_G = \min_{v_i \in V} d^G(v_i)$.

Two paths between two distinct vertices v_i and v_j are called *edge-disjoint* if they do not share any edges. The *local connectivity* $c_G(\{v_i, v_j\}) = c_G(\{v_j, v_i\})$ between two distinct vertices v_i and v_j of a graph (V, A) is the maximum number of pairwise edge-disjoint paths in (V, A).

Proposition 1. *Given a graph $G = (V, A)$ and the associated connection game (N, v), a player $i \in N, i = \{v_s, v_t\}$, is critical of order k for N if and only if the local connectivity of the vertices v_s and v_t is k.*

Proof. It is a direct consequence of Menger's theorem [7], stating that for any undirected graph $G = (V, A)$ and any pair of vertices v_i and v_j in V, the size of the minimum edge cut of v_i and v_j is equal to the number of pairwise disjoint paths from v_i to v_j, and by the consideration made in Sect. 4 about the use of the Ford-Fulkerson algorithm to compute the criticality of a player.   □

Proposition 1 suggests that computing the local connectivity of an edge of graph is equivalent to compute the criticality of this edge in the associated connection game. In general, to compute the local connectivity of all pairs of vertices need $\mathcal{O}(|A|)$ iterations Ford-Fulkerson algorithm, which can be solved in $\mathcal{O}(|V| \times |A|^2)$, so the complexity of such algorithm is $\mathcal{O}(|V| \times |A|^3)$. However, in some cases, the computation of the local connectivity of the edges is easier. First, it is immediate to notice that $c_G(\{v_i, v_j\}) \leq \min\{d^G(v_i), d^G(v_j)\}$, for any graph $G = (V, A)$ and all $v_i, v_j \in V$. In particular, two distinct vertices are said *maximally locally connected* when $c_G(\{v_i, v_j\}) = \min\{d^G(v_i), d^G(v_j)\}$.

Example 6. *As a simple example of a graph in which not all pairs of vertices are maximally locally connected, consider the graph of Example 5. In fact, we have $c_G(\{v_1, v_2\}) = c_G(\{v_3, v_4\}) = 1 = \min\{1, 2\}$, but $c_G(\{v_2, v_3\}) = 1 \neq \min\{2, 2\}$.*

For special classes of graphs, it is possible to find sufficient conditions to guarantee that any two vertices are maximally locally connected. For instance, in [9] it is shown that if $G = (V, A)$ is p-partite and

$$|V| \leq \delta_G \frac{2p - 1}{2p - 3} \tag{2}$$

then all pairs of vertices are maximally locally connected.

Another example of family of graphs where the vertices are maximally locally connected, is the class of diamond-free graphs. A *diamond graph* is an undirected graph with four vertices and five edges. A graph is *diamond-free* if it has no diamond graph as (not necessarily induced) subgraph. In [6], it is proved that if a diamond-free graph $G = (V, A)$ is such that $\delta_G \geq 3$ and

$$|V| \leq 4\delta_G - 1 \tag{3}$$

then all pairs of vertices are maximally locally connected. The following example shows a graph which satisfies such conditions.

Example 7. *Consider the graph of Example 3. Notice that it is diamond-free. Moreover, $|V| = 6 \leq 7 = 4\delta_G - 1$. On the hand, $\delta_G < 3$, so we cannot directly conclude that all pairs of nodes are maximally locally connected (and, actually, it is not the case, as $c_G(\{v_4, v_5\}) = 1 \neq \min\{3, 3\}$).*

Example 8. *Consider the following graph.*

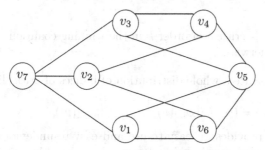

Notice that it is diamond-free and $\delta_G = 3$. Moreover, $|V| = 7 \leq 11 = 4\delta_G - 1$. Then one can easily calculate the order of criticality of each edge for coalition N in the associated connection game (N, v), which is 3 for any edge, since $c_G(\{v_i, v_j\}) = \min\{d^G(v_i), d^G(v_j)\} = 3$ for any pair of vertices v_i and v_j.

The class of diamond-free graphs is particularly interesting because if a graph $G = (V, A)$ is diamond-free, then all the induced subgraphs $G_{|M}$, with $M \subseteq A$, are diamond-free as well, and condition (3) on $G_{|M}$ together with the condition $\delta_{G_{|M}} \geq 3$ are sufficient to guarantee that all pairs of nodes in the subgraph $G_{|M}$ are maximally locally connected. In this case, in the associated connection game, for any $M \subseteq A$, an edge $\{v_s, v_t\} \in M$ is critical of order k for coalition M, if and only if $G_{|M}$ is connected and $k = \min\{|M \cap A_s^G|, |M \cap A_t^G|\}$.

6   Criticality Indices

In [4], we introduced two indices for measuring the order of criticality of an agent, based on the Shapley-Shubik index [8] and on the Banzhaf index [2], respectively.

Given a simple game v, let us start with the Shapley-Shubik index, $\phi(v)$ that is defined as the average number of times a player is critical w.r.t. a coalition when players enter the coalition in a random order. Formally, for any $i \in N$:

$$\phi_i(v) = \frac{1}{n!} \sum_{\pi \in \Pi} \sigma_i(\pi) \tag{4}$$

where Π is the class of all permutations of N and

$$\sigma_i(\pi) = \begin{cases} 1 \text{ if player } i \text{ is critical in } P_\pi^i \cup \{i\} \\ 0 \text{ otherwise} \end{cases}$$

where P_π^i is the set of players in N that precede i in the order π.

It is possible to introduce a similar index for measuring for any player $i \in N$ his power in being critical of order $k = 1, 2, \ldots, n - m + 2$ as:

$$\phi_{i,k}(v) = \sum_{S \not\ni i} \frac{|S|!(n - |S| - 1)!}{n!} \, dc_k(i, S \cup \{i\}) \tag{5}$$

where

$$dc_k(i, M) = \begin{cases} 1 \text{ if } i \text{ is critical of order } k \text{ in the winning coalition } M \\ 0 \text{ otherwise.} \end{cases} \tag{6}$$

It is possible to associate a whole distribution of indices of criticality to any player $i \in N$:

$$\Phi_i(v) = (\phi_{i,1}(v) \, \phi_{i,2}(v), \ldots, \phi_{i,n-m+2}(v))$$

An index of power provides an easy-to-use and easy-to-understand tool to compare the capability of players in being decisive, i.e. critical in forming a winning coalition. This notion may be extended by including higher order of criticality, so it is necessary to compare distributions of indices. Dall'Aglio, Fragnelli and Moretti (2019) introduce the Collective Shapley-Shubik (CSS) power index through all orders of criticality for player $i \in N$ defined as

$$\bar{\Phi}_i(v) = \frac{\sum_{h=1}^{n-m+2} \phi_{i,h}(v) h^{-1}}{\sum_{h=1}^{n-m+2} h^{-1}} \tag{7}$$

The Shapley-Shubik index of power is naturally paired with the (non-normalized) Bahnzaf index that counts the proportion of times that a certain player is critical w.r.t. any coalition that includes him. The Bahnzaf index of criticality for any player $i \in N$ for any order $k = 1, 2, \ldots, n - m + 2$ is:

$$\beta_{i,k}(v) = \sum_{S \not\ni i} \frac{dc_k(i, S \cup \{i\})}{2^{n-1}} \tag{8}$$

that leads to the distribution of indices:

$$B_i(v) = (\beta_{i,1}(v) \, \beta_{i,2}(v), \ldots, \beta_{i,n-m+2}(v))$$

Similarly, it is possible to introduce the Collective Banzhaf (CB) power index through all orders of criticality for any player $i \in N$

$$\bar{B}_i(v) = \frac{\sum_{h=1}^{n-m+2} \beta_{i,h}(v) h^{-1}}{\sum_{h=1}^{n-m+2} h^{-1}} \tag{9}$$

The indices CSS and CB can be interpreted as centrality measures of the edges for preserving the connection of the graph.

Example 9. *Consider the following graph*

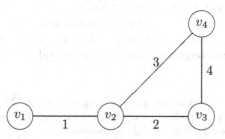

The winning coalitions are $\{1,2,3\}$, $\{1,2,4\}$, $\{1,3,4\}$, $\{1,2,3,4\}$; *in the three coalitions of cardinality 3, the three players are critical of order 1, while in the coalition of cardinality 4 player 1 is critical of order 1 and the other players are critical of order 2.*

$$\Phi_1(v) = \left(\frac{42}{42}, 0\right) \qquad\qquad B_1(v) = \left(\frac{4}{4}, 0\right)$$

$$\Phi_i(v) = \left(\frac{12}{42}, \frac{24}{42}\right), \ i = 2,3,4 \qquad B_i(v) = \left(\frac{2}{4}, \frac{1}{4}\right), \ i = 2,3,4$$

$$\bar{\Phi}(v) = \left(1, \frac{5}{7}, \frac{5}{7}, \frac{5}{7}\right) \qquad\qquad \bar{B}(v) = \left(1, \frac{5}{8}, \frac{5}{8}, \frac{5}{8}\right)$$

As it is intuitive, player 1 is the most important one for preserving the connection of the graph.

One can argue that the computation of criticality indices is quite hard, as it requires the calculation of terms $dc_k(i, M)$, whose values depend on a "comprehensive exploration" of all subgraphs G_M, for each $i \in N$ and $M \subseteq N$. As discussed at the end of the previous section, however, for specific classes of graphs like the diamond-free graphs, the calculation of these terms can be simplified, at least for those coalitions M such that $\delta_{G_M} \geq 3$. In fact, if G is a diamond-free graph, for each $M \subseteq N$ such that G_M is connected and $\delta_{G_M} \geq 3$, we have that

$$dc_k(i, M) = \begin{cases} 1 \text{ if } k = \min\{|M \cap A_s^G|, |M \cap A_t^G|\}, \\ 0 \text{ otherwise.} \end{cases} \qquad (10)$$

7   Concluding Remarks

In this paper, we have considered the class of graph connection games, and investigated the order of criticality of the players in these games. We have shown that the famous algorithm of Ford and Fulkerson [5] can be used to efficiently compute the order of criticality of players for a given coalition without considering the entire characteristic function of the game, and just looking at the structure of the graph. The relation between the notion of order of criticality and some well-known concepts of local connectivity has also been studied, and for some special classes of graphs we have noticed that the order of criticality of an edge can be easily computed looking at the degree of its extreme vertices. For future research, possible developments of the work are in the direction of identifying other classes of graphs where the computation of the order of criticality of the players is simple, and of understanding how the information about different orders of criticality can be summarized into a general definition of centrality or connectivity for edges.

Acknowledgments. The authors gratefully acknowledge the participants to the workshop "Quantitative methods of group decision making" held at the Wroclaw School of Banking in November 2018 for useful discussions.

The authors gratefully acknowledge the two anonymous reviewers for their useful comments and suggestions that allowed to improve the paper.

References

1. Aziz, H., Lachish, O., Paterson, M., Savani, R.: Power indices in spanning connectivity games. In: Goldberg, A.V., Zhou, Y. (eds.) AAIM 2009. LNCS, vol. 5564, pp. 55–67. Springer, Heidelberg (2009). https://doi.org/10.1007/978-3-642-02158-9_7
2. Banzhaf, J.F.: Weighted voting doesn't work: a mathematical analysis. Rutgers Law Rev. **19**, 317–343 (1965)
3. Dall'Aglio, M., Fragnelli, V., Moretti, S.: Orders of criticality in voting games. Oper. Res. Decis. **26**, 53–67 (2016)
4. Dall'Aglio, M., Fragnelli, V., Moretti, S.: Indices of criticality in simple games. Int. Game Theory Rev. **21**(01), 1–21 (2019)
5. Ford, L.R., Fulkerson, D.R.: Flows in Networks. Princeton University Press, Princeton (1962)
6. Holtkamp, A.: Maximal local edge-connectivity of diamond-free graphs. Australas. J. Comb. **49**, 153–158 (2011)
7. Menger, K.: Zur allgemeinen Kurventheorie. Fund. Math. **10**, 96–115 (1927)
8. Shapley, L.S., Shubik, M.: A method for evaluating the distribution of power in a committee system. Am. Polit. Sci. Rev. **48**, 787–792 (1954)
9. Volkmann, L.: On local connectivity of graphs. Appl. Math. Lett. **21**, 63–66 (2008)

The Capacity of Companies to Create an Early Warning System for Unexpected Events – An Explorative Study

Johannes (Joost) Platje[(✉)] [ID]

WSB University in Wrocław, ul. Fabryczna 29–31, 53-609 Wrocław, Poland
johannes.platje@wsb.wroclaw.pl

Abstract. This paper presents a discussion of determinants of the capacity of companies to deal with unexpected events and an approach to the creation of a company's Early Warning System. Capacity determinants discussed include: lack of functional stupidity, paradigms, general trust and awareness of fragility indicators. The results of research based on an explorative questionnaire are presented for two small Swiss and German companies. The working hypothesis for the research is that flatter organizational structures possess higher capacity to create an Early Warning System than more hierarchical organizational structures. There is some weak evidence confirming this hypothesis.

Keywords: Early Warning System · Black Swans · Knowledge management

1 Introduction

Efficiency is a main paradigm in management, and has resulted in the development of economies of scale, increasing production and wealth, as well as competitive advantages for companies seen as innovative from a managerial perspective (Hamel 2007). However, due among other factors, to increasing complexity and lengthening logistic chains at a global scale, more and more side effects can be expected resulting from innovations which implement various efficiencies (Sterman 2000; Mandelbrot and Hudson 2008; Harford 2011; Taleb 2012; Casti 2013). The moment these side effects become non-linear threats to the existence of business enterprises, it becomes more important to rethink the traditional management paradigm, and create something which may be called an Early Warning System (EWS) for low probability, high impact events that threaten the functioning or even the existence of a company (Posner 2010; Bertoncel et al. 2018).

This means that this focus on efficiency needs to be reshaped. This way of thinking, a powerful tool in initially assessing any type of venture, can be found in any introduction to economics textbook and is based on two principles: 1. There is no such a thing as a free lunch; 2. If the lunch is free, someone else pays (directly or in the form of negative externalities). However, a well-known issue identified by system theorists such as Sterman (2000), or scientists such as Benoit Mandelbrot (Mandelbrot and Hudson 2008) and Nassim Taleb (2007; 2012), is that the current models exclude many external events which should also be included in them. Markets, like life itself, are

© Springer-Verlag GmbH Germany, part of Springer Nature 2019
N. T. Nguyen et al. (Eds.): TCCI XXXIV, LNCS 11890, pp. 47–62, 2019.
https://doi.org/10.1007/978-3-662-60555-4_4

more random, volatile and full of surprises than is assumed in many economic theories (Mandelbrot and Hudson 2008). Putting it differently, even when we think we know all the costs, a surprise is very likely to appear. As our knowledge is limited in a complex world, we can assume that lack of evidence of costs is not the same as evidence of lack of costs (Taleb 2012). In this context, the following question appears: will the surprise result in opportunities for our business or threats?

The focus in this article is on negative surprises. With such surprises, the issue is whether they cause reversible, manageable problems, or potentially irreversible, non-linear threats to the very existence of the company. In the last case, standard risk management may not apply, and the precautionary principle becomes more important in decision-making (Taleb et al. 2014). In order for a company to prevent small probability, non-linear threats to itself, an EWS may be a method to ensure these issues are incorporated in the managerial process.

In this paper, first, the theoretical elements of the capacity of a company to create an EWS are presented. Then, the methodology used for empirical research is discussed together with the research results of a case study of two small companies in Germany and Switzerland. In the conclusion, implications for future research are presented. The working hypothesis for the research is that flatter organizational structures possess higher capacity to create an EWS than more hierarchical organizational structures. The latter can be expected to be more fragile, meaning that the consequences of unexpected events can be expected to be larger.

2   The Capacity for Creating an Early Warning System

An Early Warning System (EWS) should enable the capturing and identification of so-called early warning signs, which can be defined as "an observation, a signal, a message or some other item that is or can be seen as an expression, an indication, a proof, or sign of the existence of some future positive or negative issue. It is a signal, omen, or indication of future developments (Nikander 2002, 49)." Maybe one of the most important aims of an EWS is to reduce ignorance of the threats of unexpected surprises (also called Black Swans or unknown unknowns (Taleb 2007)). In order to deal with or prepare for potential threats, one needs to be aware of them (Taleb 2012). Thus, even when it is not possible to imagine what can happen, one should be aware that the unimaginable may happen. In this context, an EWS supports an environment prepared to react to and deal with "unknown unknowns". These "unknown unknowns" may in reality turn out to be "unknown knowns", as there are often some weak signals available, or information is available among stakeholders within or outside the organization. The EWS can function thus, as a kind of "smoke detector", which should include a cognitive-behavioural managerial element because formal models may miss weak signals (Bertoncel et al. 2018). "Early warning systems serve as a key management tool for anticipating potential disasters or other negative events (Trzeciak and Rivers 2003)," (quote from Bertoncel et al. (2018, 407)) in this way helping to identify, screen and appraise warning signs, and responding to them (Bertoncel et al. 2018, 412).

The focus in this paper is on elements influencing the capacity to create an EWS and, based on literature review (Taleb 2007, 2012; Sterman 2000; Meadows 1999;

Gladwin et al. 1995; Alvesson and Spicer 2012; Mandelbrot and Hudson 2008), four elements were chosen as a start point. The questions asked in the questionnaire reflect important elements of theoretical notions, which were defined and discussed during 3 small workshops with entrepreneurs in Germany, as well as 3 small workshops in Poland which took place in the period 2016–2018. The theory presented in this paper will be further developed in future cooperation with businesses.

An EWS is based on a number of fragility indicators for which a business specific questionnaire can be helpful in their identification and development. Other elements in the EWS should provide the ability/capability to look further/beyond direct evidence in order to prevent decisions that can have unexpected, negative or disastrous impact.

The capacity for companies to create an early warning system for threats to business sustainability is, among other things, determined by: 1. The existence and awareness of "fragility indicators"; 2. The level of functional stupidity (Alvesson and Spicer 2012); 3. The level of general trust (e.g., Raiser, 1997, 1999); 4. The worldview or paradigm (e.g., Meadows 1998, 1999; Sterman 2000). The meaning of these notions as well as their impact on the capacity to create an EWS are presented in Table 1.

Functional stupidity embraces different elements that hamper the flow of information within a company as well as with external stakeholders and it may be expected that this issue becomes more important in more hierarchical organizations. Williamson (1998) argues that as managerial transaction costs increase, efficient solutions for the coordination of production within the company are reduced. According to theories of asymmetric information (Akerlof 1970; Furubotn and Richter 1997), lower levels in an organizational hierarchy tend to possess more information regarding production processes than the senior levels. Therefore, management needs to create conditions for ensuring an appropriate information flow. Management support for an EWS is essential, as "[t]he propensity to disclaim disconcerting facts increases as one moves up the hierarchy [as, among other things,] corporate leaders are often not close enough to the bleeding edge of change (Hamel 2007, 45)." In particular companies with a successful management approach may forget that the basic principle of management is change so when the world changes, management needs to adapt especially as changes in the world tend to become faster and more complex (e.g., Taleb 2012), this implies that successful management needs to be prepared for innovation in management strategy as well as in organizational structure to remain successful in the long run. Thus, appropriate management provides the basis for efficient production as well as for a successful innovation strategy (Hamel 2007).

Functional stupidity can be defined as "an absence of reflexivity, a refusal to use intellectual capacities in other than myopic ways, and avoidance of justifications (Alvesson and Spicer 2012, 1188)." Following Alvesson and Spicer (2012), it embraces reflexivity (the ability and willingness to discuss and question existing knowledge, rules, norms of behaviour, etc. (Alvesson and Sköldberg 2009)), justification (reasons and explanations for decisions are provided (see Boltanski and Thévenot 2006)), and substantive reasoning (this lacks when focus is on narrow or short-term aims, and broad perspectives are not considered).

The worldview considered for research in this area is the techno-centric paradigm. This consists of a belief that technology and economic growth can solve all different types of problems (Gladwin et al. 1995). Believing all problems can be solved by

Table 1. Determinants of the capacity to create an Early Warning System

Theory	Definition	Possible effect
a. Lack of awareness of fragility issues	Fragility – something that can destroy the organization or system. Related to the Pareto Principle (20% of something causes 80% of effects; 4% of .. causes 64% of effects; 0.8% of … causes 51.2% of effects)	Increases the probability of "unknown knowns" threatening the existence of the organization. Reduces the capacity to create an EWS
b. Functional stupidity	The lack of capacity or willingness to use and apply knowledge (Alvesson and Spicer 2012) and to deal with uncertainty as well as small probability, high impact events in decision making	No learning from mistakes (related to a lack of awareness of fragility issues). Likely to increase the number of "unknown knowns" and ignorance. Reduces the capacity to create an EWS
c. Lack of general trust	Lack of trust (or distrust) in people we do not know directly	Lack of openness to ideas and solutions from people/organizations from outside a closed group. Lack of openness to information and knowledge, potentially increasing the number of "unknown knowns" and persisting the existence of ignorance. Reduces the capacity to create an EWS
d. Adherence to the technocentric paradigm	Mindset – innovation, technology and growth solve all the problems associated with sustainable development (Gladwin et al. 1995)	Lack of attention to small probability, but potentially disastrous effects (related to a. Lack of awareness of fragility issues) Lack of openness to other paradigms may lead to persistence of ignorance. Reduces the capacity to create an EWS

Source: (Platje et al. 2019), based on (Alvesson and Spicer 2012), (Gladwin et al. 1995), (Taleb 2012), (Mandelbrot and Hudson 2008).

engineering and the use of more resources is likely to lead to neglect of events that can cause significant, irreversible damage to the company.

A high level of general trust (trust in people we do not know personally) facilitates cooperation with people with new ideas. It makes the organization more adaptively efficient (Raiser 1997; 1999). Trust between stakeholders within the company is important as it may reduce barriers to flows of information crucial from the point of view of EWS. General trust may support openness to weak signals from outside the company. Furthermore, when there is a lack of general trust as well as a lack of trust in suppliers, customers and people working in the company, this may create difficulties in finding cooperative solutions requiring quick reaction in the event of unexpected events.

3 Methodology

Based on the literature study discussed above, the theoretical framework presented earlier in this paper was created and a resultant questionnaire was developed. The questionnaire was the topic of discussion at 3 workshops with entrepreneurs in Magdeburg (Germany) and 3 workshops in Wrocław (Poland) which took place between 2016 and 2018. During the workshops, after questionnaire completion, the sense of each of the questions was discussed in a group as well as which questions could be eliminated or what topics were potentially missing.

In order to test the practical relevance of the questionnaire, a German and a Swiss producer of household appliances was contacted and asked for agreement to the questionnaires being completed anonymously within their company. The results and the final paper were reviewed and consulted with the CEO and the owner of the respective companies The Swiss company is a family company, with the German company being owned in-turn by the Swiss company. The questionnaire was filled out by all employees and owners of the companies during a special meeting in Spring 2018. All employees and managers of the company were asked to complete the questionnaire, as at different levels of an organization people may possess different knowledge about potential fragilities.

The purpose and meaning of the questionnaire was clearly explained, participation was voluntary and confidentiality was assured by the CEO and the owner of both companies. All questionnaires were sent to and analysed by the author of this paper. The general results were discussed with the CEO of the German company and owner of the Swiss company. Feedback from their side was provided in October 2018. The Swiss company employs 41 people, of which 38 filled out the questionnaire. The German company employs 54 people, of which 46 filled out the questionnaire.

The main hypothesis of the research is that functional stupidity is correlated with a hierarchical organizational structure. An important reason for this is that bureaucratic processes and problems of asymmetric information tend to increase in more hierarchical structures (Hamel 2007; Furubotn and Richer 1997). Below, the general results are presented, based on simple statistical analysis and the discussion during the meetings with the owner and with the CEO. Differences between the German and Swiss company were tested using the R program. As both companies expressed interest in future cooperation, the questionnaire is planned to be repeated in 2019. The results should be interpreted with care – as they provide just a general picture and that picture is only a one moment "snap-shot". For this reason, the research should be repeated. The questionnaire consisted of 53 statements concerning the theoretical framework on the capacity to create an EWS (presented in Sect. 2 – Tables 2, 3, 4, 5, 6, 7 and 8). Also included were some dummy questions. The results on the two companies' capacity to create an EWS are presented and discussed below. Respondents could answer on a 1–5 scale anything from "completely disagree" (1) to "completely agree" (5), 3 being "neither disagree nor agree". Also, options for: "don't know", "I do not possess information on this issue", as well as "confidential information" were provided.

Important is the level of (dis)agreement with different statements, providing information on lack of capacity to create an EWS (i.e., lack of identification of a specific problem), enabling reflection on the possible implications of this lack of capacity. The following two questions were asked at the end of the questionnaire: (a) "Please identify maximum 5 questions that, according to you, can be eliminated from the questionnaire. Please give the numbers of these questions." (b) "Please identify maximum 5 questions that, according to you, are very relevant. Please give the numbers of these questions." While question (b) allows for discussing issues which according to the respondents are important, question (a) requires deeper elaboration as it may on the one hand, allow for the elimination of questions regarded as not useful for future versions of the questionnaire. On the other-hand, the questions identified for elimination may also indicate areas of potential fragility. The reason for this is that while a question may seem not useful for the respondent, this perceived lack of usefulness does not necessarily mean topic uselessness (Taleb 2012).

4   Data and Discussion

The results of the questionnaires in the German and Swiss companies are presented in Tables 2, 3, 4, 5, 6, 7 and 8 with calculations for the mean, difference between means (two-tailed t-test), variance and correlation (Pearson's rho). This allows for an initial assessment of the capacity to create an EWS as well as identification of fragilities, and analysis of the main differences between the two companies. A mean of three (either within or between companies) implies that, on average, the respondents neither agree nor disagree. A mean below 3 means that respondents tend to disagree, while over 3 implies that they rather agree.

Table 2 presents data on questions regarding the perceived state of reflexivity and opinions about it in the two companies where it seems that the level of reflexivity in the Swiss company is higher than in the German company. The two tailed t-test shows a significant difference in the mean for question 2 regarding the possibility of criticizing management (p 0.0171) and question 5 whether this issue is a problem (p 0.0286). In the German company, the state of reflexivity requires deeper research, as the mean points indicate respondents neither agree nor disagree regarding criticism of management, and open discussion of changes in rules. Opinions on these issues show that respondents think that a lack of reflexivity is problematic. In the case of the German company, weak correlation has been observed between not talking about mistakes in the company (question (1)) and an authoritarian management style (Table 5) with a Pearson's r = 0.4679999, p-value = 0.009106. Weak correlation was also observed for the German company between question 9 (Table 3) – that respondents think it is not a problem when the management of a company does not provide reasons and explanations for their decisions and a hierarchical management structure - Pearson's r = 0.4887966, p-value = 0.009677. For the Swiss company, no correlations with other questions have been observed.

Table 2. Functional stupidity – state of and opinions about reflexivity

Questions	DE mean	SW Mean	DE St. dev.	SW St. dev.
(1) We do not talk about mistakes in our company	2.585365854	2.243243243	1.139640508	0.983344785
(2) It is possible to doubt/criticize management ideas/decisions in our company	3.088235294	3.647058824	0.933148696	0.94971616
(3) Changes in the rules are openly discussed in our company	3.047619048	3.333333333	0.986553027	0.956182887
(4) It is not a problem when mistakes are not discussed in the company	2.027027027	1.696969697	0.985632524	0.847232572
(5) It is not a problem when it is impossible to express doubts and criticize management decisions in a company.	2.363636364	1.878787879	0.976095755	0.599873724
(6) It is not a problem when changes in the rules in a company are not openly discussed	2.216216216	1.914285714	1.031049496	0.701738537

DE – German company. SW – Swiss company. A 1–5 scale was used, from "completely disagree" (1) to "completely agree" (5), 3 being "neither disagree nor agree".

The differences in the mean can be explained by the number of respondents (completely) disagreeing. With statement 1 "We do not talk about mistakes in our company", 55% of the total amount of respondents (including the respondents not giving an opinion) of the German company (completely) disagree, while the percentage is 70% for the Swiss company. The owner of the Swiss company argued that the reasons may be that Switzerland is characterized by flatter managerial structures with a high degree of competitive pressure on Swiss companies due to high labour costs. The employees see a greater need for learning from mistakes, as they are closer to the company management. With statement 2, "It is possible to doubt/criticize management ideas/decisions in our company", 37% of the German respondents (completely) agree, while this percentage is 67% for the Swiss company. With statement 3, "Changes in the rules are openly discussed in our company", respectively 43% and 50% (completely) agree.

The questions regarding the state of reflexivity were considered to be important by 1 German respondent and 9 Swiss respondents. The questions regarding opinions about reflexivity were considered to be important by 6 German respondents and 11 Swiss respondents, while respectively 4 and 8 respondents thought these questions could be deleted from the questionnaire. In particular questions 4 (8 respondents consider it important) and question 6 (5 respondents consider it important) were considered to be important.

The level of justification (Table 3) seems not to differ between the Swiss company and the German company. However, regarding the lack of providing reasons and explanations for managerial decisions (question 9), the German respondents see less problems with this aspect than the Swiss ones (respectively 32% and 45% (completely) disagree). The two tail t-test shows a significant difference between the means (p 0.04). This issue may need attention, as 3 German respondents think this question is irrelevant. Among the Swiss respondents, 4 consider this question important, while only 1 German respondent has the same opinion. As mentioned earlier, responses to this question are correlated with the hierarchical management structure in the German company.

Table 3. Functional stupidity – state of and opinions about justification

Questions	DE mean	SW Mean	DE St. dev.	SW St. dev.
(7) The company management often provides reasons and explanations for their decisions	3.43902439	3.378378378	0.895789609	0.981816665
(8) People in the company provide, when necessary, feedback to other people in the company	3.648648649	3.777777778	0.823819567	0.680802514
(9) It is not a problem when the management of a company does not provide reasons and explanations for their decisions	2.4	1.971428571	1.005865153	0.663578281
(10) It is not a problem when people in the company do not provide feedback to other people when necessary	1.818181818	1.727272727	0.882274952	0.761278828

DE – German company. SW – Swiss company. A 1–5 scale was used, from "completely disagree" (1) to "completely agree" (5), 3 being "neither disagree nor agree".

The state of substantive reasoning seems fairly positive. About 60% of the Swiss and German respondents (completely) agree with statement 12 "Things that almost went wrong are discussed together with the lessons learned". Learning from these experiences is an important element of an EWS (Bertoncel et al. 2018) and this question is considered to be important by 6 respondents. Question 11 is considered to be important by 5 respondents. Regarding statement 13, 3 respondents think this issue is very important, while 2 respondents think it could be removed from the questionnaire. Regarding statement 14 on ignorance of low probability threats, 14% of the German and 8% of the Swiss respondents think this case applies, while 2 respondents think this question is very important and 1 respondent thinks this question could be removed. This issue may need attention in the company management strategy, as both

Table 4. Functional stupidity – state of and opinions about substantive reasoning

Questions	DE mean	SW Mean	DE St. dev.	SW St. dev.
(11) When making a mistake in our company, there is a positive atmosphere to find a solution	3.325	3.675675676	0.91672494	0.851601565
(12) Things that almost went wrong are discussed and lessons have been learned	3.348837209	3.513513514	0.973059019	0.931594261
(13) Our company ignores threats to its existence which are difficult to quantify	2.379310345	2.416666667	0.621851676	0.880546602
(14) Our company ignores low probability threats	2.666666667	2.4	0.877058019	0.816496581
(15) It is not a problem when a company ignores threats to its existence which are difficult to quantify	2.090909091	1.636363636	1.011299794	0.65279121
(16) It is not a problem when a company ignores low probability threats	2.242424242	1.909090909	0.867118181	0.678400525

DE – German company. SW – Swiss company. A 1–5 scale was used, from "completely disagree" (1) to "completely agree" (5), 3 being "neither disagree nor agree".

of the companies are dependent on one critical element for which no cheap quality approved substitute is available. The Swiss company has experience with fragilities as, due to a fire a 7 year stock of this element was destroyed. The company managed to continue production for a couple of weeks by applying emergency measures, and has now (somewhat) diversified the risk by having stock at two locations (this may thus be a case of an "unknown known"). The company also started to look for a second supplier however, it took some time before the quality of the product from the second supplier was approved. This is an example of quality requirements which are transaction-specific (see Williamson 1998) and thus create fragility. After such a disastrous event, people thought it was obvious to put crucial stock in more than one place. However, when no accidents occur, most managers will just not think about the small probability that everything will go wrong. A reason for this is that people often tend to forget about the obvious (Beck 2017) whereas disasters are often caused by a chain of unlikely mistakes (Harford 2011). Another reason why small probability events with a high impact need managerial attention, is that 4 respondents think statement 15 can be removed, while 3 think it is important. The two tailed t-test shows a significant difference between the means (p 0.034). Regarding question 16, 3 respondents think this question can be removed, while 1 respondent thinks it is very important. In the German company, 4 respondents agree with statements 15 and 16, whereas only 1 in the Swiss company. While 4 out of 46 respondents may seem a small part of the total, the issue with small probability high impact events is that in order to create an EWS, information and weak signals regarding potential threats should flow from all parts of the organization to the management (Bertoncel et al. 2018).

Table 5. Management style

Question	DE mean	SW Mean	DE St. dev.	SW St. dev.
(17) The management style of our company is authoritarian	3.090909091	2.264705882	0.97991187	1.109431291

DE – German company. SW – Swiss company. A 1-5 scale was used, from "completely disagree" (1) to "completely agree" (5), 3 being "neither disagree nor agree".

The data in Table 5 show that the German company is considered to have a more authoritarian management style than its Swiss counterpart with the two tailed t-test showing a significant difference between the means (p 0.00196). In Germany, 20% of the respondents (completely) disagree, while 30% (completely) agree. Of the respondents in its Swiss counterpart, 67% (completely) disagree, while 19% (completely) agree. A weak inverse correlation (Spearman's r - −0.397203; p-value = 0.04021) between an authoritarian management style, and the lack of time to introduce and manage these changes, has been observed for the German company (question 18). A similar weak inverse correlation has been observed between management style, and dependency on a few suppliers (question 19, Table 6, Pearson's r - −0.4026499, p-value = 0.03732). For the Swiss company, no correlation has been observed.

In Table 6, different types of fragilities are presented. The data indicate that the German company is more dependent than the Swiss company on one or several good employees (question 19) and managers (question 20). The two tailed t-test shows a significant difference between the means (respectively p 0.00072 and p 0.001). About 45% of the German respondents (completely) agree that their company is dependent on a few good employees, while 21% (completely) disagree. The percentages are 18% and 58% respectively for the Swiss company. Furthermore, the German respondents (including the CEO) think that innovations make the company reliant on high skilled, difficult to find employees (question 25), strengthening reliance on a small number of good employees. The two tailed t-test shows a significant difference between the means (p 8.29456E-06) with the means for the answers regarding employees in the German company (question 19 and 25) in the range of 3.4–3.5 (which can be interpreted as a kind of "rather agree").

Table 6. Different types of fragilities – state in the company

Questions	DE mean	SW Mean	DE St. dev.	SW St. dev.
(18) In general, there are too many changes in our company, and too little time to introduce and manage these changes	2.848484848	2.735294118	1.003780732	0.863706724
(19) Our company is dependent on one or a few good employees	3.414634146	2.516129032	1.071811778	1.060533443
(20) Our company is dependent on one or a few good managers	3.171428571	2.35483871	1.070615938	0.877435167
(21) Our company is dependent on one or a few suppliers	2.625	2.818181818	1.008032258	0.957992128
(22) If necessary, it would be easy for our company to find new suppliers	3.064516129	3.05	0.813858459	0.686332741
(23) If necessary, it would be easy for our company to find new customers	2.90625	2.615384615	0.92838309	0.803836952
(24) Our company is dependent on one or a few customers	2.393939394	2.24137931	0.788170109	0.786273871
(25) The innovations of our company increase the reliance of high skilled and difficult to find employees	3.533333333	2.333333333	1.04166092	0.637022057
(26) The innovations of our company have made it more reliant on one or two suppliers	2.6	2.866666667	0.816496581	0.915475416
(27) Innovations in our companies have made management more complex	2.62962963	2.470588235	0.88353086	0.799816155

DE – German company. SW – Swiss company. A 1-5 scale was used, from "completely disagree" (1) to "completely agree" (5), 3 being "neither disagree nor agree".

In the subsequent questionnaire results discussion with the Swiss owner and the German CEO, the reliance of both companies on a single supplier for a crucial production element, for which no cheap substitute with approved quality is easily available was identified as an issue. The current supplier is relatively cheap and the quality of the element is known by both companies therefore, a change in supplier could not only affect costs, but could also create a difficult to quantify quality risk. This topic formed part of the discussion of difficult to quantify and small probability threats. Another topic discussed was that, through the years, the German company has had less employee turnover than the Swiss company. A reason discussed may be that workers of the company work in one shift until 15:30 PM, while competitors produce in three shifts. Another reason may be shorter lines in human resource management in the SME, enabling a more personal approach towards employees.

Table 7. Different types of fragilities – opinions in the company

Questions	DE mean	SW Mean	DE St. dev.	SW St. dev.
(28) It is not a problem when innovations in a company increase the reliance on high skilled, difficult to find employees	3	2.153846154	0.870988341	0.543492976
(29) It is not a problem when the innovations of a company make it reliant on one or two suppliers	2.575757576	1.851851852	1.031694693	0.533760513
(30) Stocks and buffers which seem not to be useful can be eliminated	3.125	2.925925926	1.157026222	0.997146927
(31) It is not a problem when the innovations of a company make the management more complex	2.774193548	2.538461538	1.023382542	0.859337849

DE – German company. SW – Swiss company. A 1-5 scale was used, from "completely disagree" (1) to "completely agree" (5), 3 being "neither disagree nor agree".

Differences between the German and Swiss companies can also be observed in ways of thinking. The two tailed t-test shows significant differences between the means of the answers to questions 28 (p 7.67274E-05) and 29 (p 0.001661019). The German respondents tended to think that increasing reliance on difficult to find employees and single suppliers is less problematic than for the Swiss. A weak correlation between the response that stocks and buffers that seem not to be useful and thus, can be eliminated (question 30, related to the idea that lack of evidence of usefulness is not the same as evidence of uselessness) combined with a perceived hierarchical management structure (question 17) has been observed for the German company (Pearson's r - 4092908; p-value = 0.03787). A weak inverse correlation (Pearson's r - −0.4045712; p-value = 0.04486) has been observed for the lack of perception of problems connected with increasing management complexity due to the need to support innovations (question 31) and perceived hierarchical management structure (question 17). Regarding question 31, 4 respondents thought this is an important issue, and 4 respondents thought this question could be eliminated from the questionnaire. Regarding question 31, 5 respondents thought this could also be deleted (4 Swiss, 1 German), while 4 thought it was a very important question (1 Swiss, 3 German). The results regarding stocks and buffers may be an indication that more reflection on risk management may be needed. Finally, increased complexity in management due to innovations may be an issue for deeper reflection in particular, in the German company.

Table 8. Trust and worldviews

Questions	DE mean	SW Mean	DE St. dev.	SW St. dev.
(32) In general, people can be trusted	3.186046512	3.486486486	0.95757168	0.803520781
(33) In general, people in our company can be trusted	3.441860465	3.75	0.795892	0.76997217
(34) In general, our company's customers can be trusted	3.111111111	3.571428571	0.69798244	0.597614305
(35) In general, our company's suppliers can be trusted	3.222222222	3.666666667	0.69798244	0.483045892
(36) Innovations and development of technology will solve problems with environmental pollution and overuse of natural resources	2.677419355	2.59375	0.70176429	1.042929342

DE – German company. SW – Swiss company. A 1–5 scale was used, from "completely disagree" (1) to "completely agree" (5), 3 being "neither disagree nor agree".

Different types of trust tend to be higher in the Swiss company than in its German counterpart (Table 8). The outcome that the Swiss employees tend to trust people, their co-workers, suppliers and customers, is important in the context of, as expressed by the owner, the strong competitive environment. This strong competitive environment, accompanied by high labour costs in Switzerland, increases the importance of internally developed human capital for innovations (see Hamel 2007). The two tailed t-test shows a significant difference between the means regarding trust in customers (question 34 - p 0.02) and trust in suppliers (question 35 - p 0.0165). A total of 12 respondents thought that the question on general trust (question 32) could be removed (4 German, 8 Swiss), while only 2 respondents thought this question was important (1 German, 1 Swiss). Interestingly, 8 respondents thought that question 33 (that people in the company can be trusted) is very important. Only 1 thought the question could be removed. This type of trust is important for information transfer within the company as well as for innovation (Hamel 2007), and is an important element of an EWS. Furthermore, 2 respondents thought that the question on trust in customers could be removed, while 3 thought the question on trust in suppliers could also be removed. The importance of different types of trust in management should also be a topic for reflection in both companies. For example, general trust facilitates cooperation with external stakeholders, and increases openness to new ideas from outside the company. Lack of awareness of the options for innovation and lack of openness to obtaining relevant information from external stakeholders, not only reduces these options, but it also reduces the capacity to catch weak signals, which is an important element of an EWS.

Regarding question 36 (innovations and development of technology will solve problems with environmental pollution and overuse of natural resources), 6 respondents thought it could be removed. In the German company, 36% (completely) disagreed with this statement, while 47% for the Swiss company. In the German company, 17% (completely) agreed, whilst this percentage is 24% for the Swiss company (including 2 managers). No correlation has been observed between this worldview and determinants of functional stupidity. A believe that technology solves all problems may lead to a neglect of small probability, high impact events. It also may lead to a neglect of fragilities which become apparent as a result of innovations. This may be due to a belief that technology will solve the new problems which will appear in the future. This issue should also be a topic for deeper reflection especially for the Swiss company.

5 Concluding Remarks

In this paper, the results of research in two companies regarding the capacity for creating an Early Warning System (EWS) to identify and deal with unexpected, low probability, high impact events was presented. The working hypothesis used as a background for the research is that flatter organizational structures possess higher capacity to create an EWS than more hierarchical ones. The latter can be expected to be more fragile, indicating that the impact of unexpected events can be expected to be greater. There is some evidence supporting this hypothesis. However, the evidence is weak, and only based on a case study of two small companies. For the companies themselves, the results may be a basis for further reflection.

It can be argued that a low level of functional stupidity is a condition for reducing ignorance among management, owners and company employees. When accompanied by trust, a process can be started to create an EWS, beginning with the fragility indicators included in the questionnaire. However, this is only a starting point for a deeper discussion hopefully leading to management involvement in developing an EWS. A next step in the research process is a set of in depth interviews with the company owners and management regarding an EWS focused on their specific business profile.

The research shows that a weak correlation was apparent between some elements of functional stupidity and the perceived authoritarian management structure in the German company. Likewise, a weak correlation was observed between some fragility issues (dependency on difficult to find employees as well as good managers) and the perceived authoritarian management structure in the German company. This is a weak corroboration of the hypothesis that fragility and functional stupidity is related to an authoritarian management structure. In addition, the German company is more dependent on one or a few good managers, which may be related to its more hierarchical structure. This raises another topic for future research - to what extent dependency on one or a few good managers can threaten the company in the event of unexpected random events, and whether this issue is related to hierarchical organization structures. Lack of managerial innovations may create a serious threat to long term organizational viability, while radical managerial innovation may lead to a long term, difficult to copy, competitive advantage (see Hamel 2007).

Issues of trust need attention, in particular within the Swiss company, as respondents seem to underestimate the importance of general trust for company development. Of course, the respondents may also have thought that they need to pick questions that should be eliminated, and therefore the question regarding general trust is one of the least directly related to the functioning of the company. However, looking beyond the visible and direct goals of a company is an element of substantive reasoning and substantive reasoning, by ensuring the inclusion of small probability, difficult to measure and high impact events into managerial practice, may be a necessary condition of developing an EWS. This is related to the awareness of potential threats (in fact a kind of worldview), which results in a process of creating an EWS that goes beyond strictly technical threats.

An issue that needs deeper elaboration is the relation between worldviews and the other determinants of the capacity to create an EWS. Theoretically. It may be that people are clever and think critically in an environment where a high level of general trust exists and fragility indicators may have been implemented. But when people believe that technology solves all problems, they can create even bigger problems with their intelligence and creativity. Among other reasons, because not enough buffers (time, money, natural resources, networks of contacts, etc.) are created, while simultaneously neglecting the need for some type of Early Reaction System as an element of an EWS that allows for rapid learning in times of extreme volatility (see Posner 2010).

References

Akerlof, G.A.: The market for "lemons": quality, uncertainty and the market mechanism. Quart. J. Econ. **84**, 488–500 (1970)

Alvesson, M., Spicer, A.: A stupidity-based theory or organizations. J. Manag. Stud. **49**(7), 1186–1220 (2012)

Alvesson, M., Sköldberg, K.: Reflexive Methodology. Sage, London (2009)

Beck, H.: Irren ist nuetzlich. Carl Hausner Verlag, Muenchen (2017)

Bertoncel, T., Erenda, I., Pejić Bach, M., Roblek, V., Meško, M.: A managerial early warning system at a smart factory: an intuitive decision-making perspective. Syst. Res. Behav. Sci. **35**, 406–416 (2018)

Boltanski, L., Thévenot, L.: On Justification. Princeton University Press, Princeton (2006)

Casti, J.L.: X-Events – Complexity Overload and the Collapse of Everything. Harper Collins Publishers, New York (2013)

Furubotn, E.G., Richter, R.: Institutions and Economic Theory - The Contributions of the New Institutional Economics. The University of Michigan Press, Ann Arbor (1997)

Gladwin, T.N., Kennelly, J.J., Krause, T.-S.: Shifting paradigms for sustainable development: implementations for management theory and research. Acad. Manag. Rev. **20**(4), 874–907 (1995)

Hamel, G.: The Future of Management. Harvard Business Review Press, Boston (2007)

Harford, T.: Adapt – Why Success Always Starts with Failure. Little, Brown, London (2011)

Mandelbrot, M., Hudson, R.L.: The (mis)behaviour of Markets. Profile Books, London (2008)

Meadows, D.: Indicators and Information Systems for Sustainable Development. The Sustainability Institute, Hartland (1998)

Meadows, D.: Leverage Points – Places to Intervene in a System. The Sustainability Institute, Hartland (1999)

Nikander, I.O.: Early warnings: a phenomenon in project management. Doctoral dissertation. Helsinki University of Technology, Department of Industrial Engineering and Management, Espoo, Finland (2002). http://lib.tkk.fi/Diss/2002/isbn9512258889/isbn9512258889.pdf

Platje, J., Will, M., Van Dam, Y.: A fragility approach to sustainability – researching effects of education. vol. ahead-of-print No. ahead-of-print (2019). https://doi.org/10.1108/IJSHE-11-2018-0212

Posner, K.: Stalking the Black Swan: Research and Decision Making in a World of Extreme Volatility. Columbia Business School Publishing (2010)

Raiser, M.: Informal institutions, social capital and economic transition: reflections on a neglected dimension. EBRD Working paper 25, London (1997)

Raiser, M.: Trust in transition, EBRD Working paper 39, London (1999)

Sterman, J.D.: Business Dynamics: System Thinking and Modelling for a Complex World. Irwin/McGraw Hill, Boston (2000)

Taleb, N.N.: The Black Swan - The Impact of the Highly Improbable. Penguin Books, London (2007)

Taleb, N.N.: Antifragile - Things that Gain from Disorder. Penguin Books, London (2012)

Taleb, N.N., Read, R., Douady, R., Norman, J., Bar-Yam, Y.: The Precautionary Principle: fragility and Black Swans from policy actions, Extreme risk initiative – NYU School of Engineering Working Paper Series (2014). https://arxiv.org/pdf/1410.5787.pdf

Trzeciak, S., Rivers, E.P.: Emergency department overcrowding in the United States: an emerging threat to patient safety and public health. Emerg. Med. J. 20(5), 402–405 (2003)

Williamson, O.E.: Transaction cost economics: how it works; where it is needed. De Economist 146(1), 23–58 (1998)

Electoral Reform and Social Choice Theory: Piecemeal Engineering and Selective Memory

Hannu Nurmi[✉]

Department of Contemporary History, Philosophy and Political Science,
University of Turku, Turku, Finland
`hnurmi@utu.fi`

Abstract. Most electoral reforms are dictated by recognized problems discovered in the existing procedures or - perhaps more often - by an attempt to consolidate power distributions. Very rarely, if ever, is the motivation derived from the social choice theory even though it deals with issues pertaining to what is possible and what is impossible to achieve by using given procedures in general. We discuss some reforms focusing particularly on a relatively recent one proposed by Eric Maskin and Amartya Sen. It differs from many of its predecessors in invoking social choice considerations in proposing a new system of electing representatives. At the same time it exemplifies the tradeoffs involved in abandoning existing systems and adopting new ones.

Keywords: Condorcet consistency · Plurality voting · Plurality with runoff · Black's method · Nanson's rule

1 Introduction

As basically all institutions, the voting procedures can be seen as problem-solving devices: their adoption starts with a recognized problem – be it one of including popular participation in a domain where decisions formerly were made by a single individual or one related to the working of an existing procedure – proceeding then to a mechanism that purportedly provides a solution to this problem. This piecemeal way of improving existing institutions may, however, lead to a paradox that carries the name of Marquis de Condorcet, the French social philosopher and science administrator of the 18th century. We shall give it a slightly unusual interpretation in the following Table 1. Consider three voting procedures—x, y and z – three performance criteria – I, II and III – and the following hypothetical configuration of procedures and criteria.

In Table 1 the procedures are ranked on each of the three performance criteria. Thus, e.g. on criterion II y performs best, z second best and x worst. Assuming that the criteria are of equal importance, it is reasonable to come up

The author thanks the referees for numerous constructive comments.

Table 1. Condorcet's paradox

Criterion I	Criterion II	Criterion III
x	y	z
y	z	x
z	x	y

with the overall conclusion that x is better than y since it is ranked higher than y on two criteria (I and III). Similarly, y can be deemed better than z since it is ranked higher than z on two criteria (I and II). One could then be led to conclude by transitivity that x is better than z as well. Yet, this is clearly not the case: z is ranked higher than x on two criteria (II and III). This is the crux of Condorcet's paradox: aggregating several rankings (i.e. complete and transitive preference relations) applying the majority principle in each pair of alternatives may lead to an endless cycle of alternatives where each replacement of an alternative can be justified by a majority rule, but at the same time no best alternative is to be found.

This is a version of the well-known money-pump. To wit, suppose the original procedure is x which is the best of the three in terms of criterion I. Assume now that an analyst or a proponent of y comes up with criterion II which is deemed more plausible than I. In terms of criterion II the ranking of the procedures is the one presented in the middle of Table 1. Now an advocate of z may take the floor and suggest criterion III where z is ranked first, x second and y third. This might be the historical sequence of events whereby criteria are considered one by one. Depending on the content of the criteria each step of the process from x to z via y may well be justified, but the overall result - i.e. when all criteria are taken into account and given a roughly equal weight – is a never-ending cycle of 'improvements'.

In the next sections we shall focus on a couple of examples of electoral reform. First, we discuss the proposal of Jean-Charles de Borda presented for the French Royal Academy in the 1770's. We then focus on a procedure designed to rectify an alleged weakness in Borda's proposal. Borda's proposal was primarily directed avoiding the main flaw of the first-past-the-post (plurality) voting. Other ways of avoiding this are discussed next. Finally, we deal with the recent proposal presented by Amartya Sen and Eric Maskin to replace the prevailing first-past-the-post system of the U.S. Congressional elections with a hybrid of two mutually incompatible procedures.

2   The First Attack on the Plurality System

The one-person-one-vote or plurality voting system is undoubtedly the most common voting procedure today. It is not only used extensively in political elections (e.g. in the U.K. parliamentary and in the U.S. elections of the members of the House of Representatives), but also in informal settings. In fact, it is the procedure people typically have in mind when saying 'let's take a vote' in informal settings. Its main virtue is simplicity, both for voters and for the persons determining the winner(s). Its outcomes also lend themselves for a straightforward interpretation: the winner is the alternative that has been voted upon by more voters than any of its contestants. But it is associated with an important shortcoming made visible by Borda in his example reproduced in Table 2 [4,11,16].

Table 2. Borda's paradox

1 voter	7 voters	7 voters	6 voters
A	A	B	C
B	C	C	B
C	B	A	A

Assuming, as Borda does, that each voter votes according to his/her preferences, alternative A wins with 8 votes against 7 (for B) and 6 (for C). And yet, A is the last-ranked alternative of a clear majority of voters, an absolute (Condorcet) loser one could say in modern terminology. As such it would be defeated in pairwise majority comparisons by all other alternatives. It seems that Borda's main motivation was to prevent such an outcome from happening by devising a system that would exclude the eventual Condorcet losers from being elected. Thus, he introduced a procedure: the method of marks or the Borda count in modern terminology. Given a profile of individual preference rankings over, say k, alternatives, the Borda count determines the Borda scores of alternatives as follows: each voter ranking an alternative first increases this alternative's score by $k - 1$ points, each voter ranking an alternative second increases the score by $k - 2$ points, etc. and each voter ranking an alternatives last increases its score by zero points. The sum of points given by all voters determines its Borda score. The collective ranking of alternatives is the same as the order of their Borda scores. Hence, it is natural to call the first ranked alternative the Borda winner. Borda showed that his method can also be implemented through exhaustive pairwise comparisons so that the score of each alternative is the sum of the votes it gets in every pairwise comparison with other alternatives.

That the Borda count really guarantees the exclusion of the eventual Condorcet loser was formally shown by P. C. F. Daunou in the beginning of the

Table 3. The strong Condorcet winner and the Borda winner can be distinct

4 voters	3 voters
A	B
B	C
C	A

19th century [11, 263–267].[1] Table 2 gives rise to another observation that came to play a central role in the emerging debate between Borda and Condorcet, viz. the possibility that the procedure investigated always elects a Condorcet winner, that is, an alternative that – on the basis of the information presented in the preference profile – defeats by a majority of votes all others in pairwise comparisons. This is often deemed an important desideratum in the theory of voting (see e.g. [5]; for more sceptical views, see [17, 20]). In Table 2 there is a Condorcet winner[2] and it is simultaneously the Borda winner. This, however, is not always the case. Indeed, one could even envision a profile with a strong Condorcet winner which, nonetheless, is not the Borda winner. Table 3 provides an illustration.

Here A is the strong Condorcet winner, but B gets the largest Borda score.

3   The Search for Condorcet Consistency

For those convinced of the desirability of Condorcet consistency – i.e. the property guaranteeing the choice of Condorcet winner when one exists – the possibility of not electing a Condorcet winner is not acceptable. One of the early writers feeling uncomfortable at the possibility of not electing a Condorcet winner was

---

[1] It is relatively straightforward to see how this conclusion is derived. To wit, suppose that there is a Condorcet loser, say x, in a profile consisting of n voters and k alternatives. This means that in each pairwise comparison, the maximum number of votes for x is strictly less than $n/2$. Hence x's Borda score is less than $(k-1) \times (n/2)$. If all alternatives had the same or smaller Borda score than x (which would make x the Borda winner), the total number of Borda scores would be no larger than: $k \times (k-1) \times n/2$. Now this upper bound is strictly less than the sum total of Borda scores in any profile, viz. $n \times (k^2 - k)/2 = k \times (k-1) \times n/2$. (The number of pairwise comparisons involving different alternatives is $k^2 - k$ with the sum of entry (i, j) and entry (j, i) being equal to the number of voters, n, for all alternatives i and j.) Therefore, in any profile there must be at least one alternative with a strictly larger Borda score than that of the Condorcet loser. Hence, the latter cannot be elected by the Borda count.

[2] A strong Condorcet winner is an alternative ranked first by more than half of the electorate. Obviously, all procedures that elect a Condorcet winner also elect a strong Condorcet winner. The converse is not true, that is, there are procedures (e.g. plurality voting) that elect the strong Condorcet winner, but not necessarily a Condorcet winner.

E. J. Nanson. He set out to design a system that would be as 'Bordaesque' as possible while at the same time satisfying Condorcet consistency [13]. The resulting procedure – known as Nanson's rule – is based on a sequence of eliminations based on the Borda scores of alternatives so that in each counting round those alternatives with the average or smaller Borda score are discarded whereupon the Borda scores are computed in the remaining set of alternatives. Eventually the winner is found after one or more rounds of eliminations.

The main purpose of Nanson's rule is to preserve the eventual Condorcet winner, while determining the winner by using in full the positional information given by the voters just like in the Borda count. The survival of the Condorcet winner in the elimination process is guaranteed by a relationship that exists between the Condorcet and Borda winner in any profile: the Condorcet winner always has a Borda score that is strictly larger than the average.[3] Hence, the Condorcet winner, whenever it exists, coincides with the Nanson winner.

A few decades after the invention of Nanson's rule, another Borda elimination procedure was proposed by Baldwin [1]. Somewhat paradoxically Baldwin – fully cognizant of Nanson's rule – presented his method as a simplification of Nanson's even though it in general requires more computing rounds than Nanson's rule. The basic difference between these two is that Baldwin's method eliminates at each counting round the alternative with the smallest Borda score. It is easy to see that, as Nanson's, also Baldwin's method is Condorcet-consistent.[4]

4   Meanwhile Other Attacks Were Launched

The possibility that an alternative elected is considered very bad (even the worst) by more than a half of the electorate has motivated other voting procedure proposals. By far the most successful of them in terms of the frequency of adoption is the plurality with runoff procedure. It is clearly aimed at securing that the winner is supported by at least half of the electorate. If this is not the case in the original profile, the runoff is organized between those two (sometimes three or more) alternatives gaining the most votes in the original profile. With only two alternatives competing on the second round and barring ties, one of these is bound to receive the support of more than half of the electorate.

This method guarantees that an alternative ranked last by a majority of voters cannot be elected. For if the winner is found in the original profile, the winner cannot be such an alternative that is ranked last by a majority of voters (since such an alternative cannot simultaneously be first-ranked by a majority). If, on the other hand, the winner is found on the second round, it cannot be the alternative that is originally last-ranked by a majority (since that majority ranks whichever of its competitors first in the second round).

---

[3] Nanson's argument to that effect is pretty similar to the one in footnote 1. It amounts to showing that the lower bound of the Borda score of a Condorcet winner is strictly larger than the average of the Borda scores.

[4] The differences and similarities of Nanson's and Baldwin's rules are discussed in [7,14,15].

Table 4. Non-monotonicity of plurality with runoff and alternative vote

8 voters	9 voters	7 voters
A	B	C
C	C	B
B	A	A

A couple of decades earlier than Nanson presented his rule, Carl Andrae of Denmark and Thomas Hare of England introduced the alternative vote method. In contradistinction to Nanson's rule, this method is based on plurality eliminations of alternatives one by one until one of them occupies a majority of first ranks. In three-alternative profiles the plurality runoff and alternative vote are equivalent, but with the presence of more than three alternatives they can end up with different choices. A common feature in both these procedures is that they never choose a candidate ranked last by a majority of voters. The reason for this in the case of plurality runoff was just stated. In the case of the alternative vote, the reason is the observation that in order to become the alternative vote winner, the alternative has to defeat by a majority at least one other, viz. the one it is confronted with in the last sub-profile that determines the winner.

5   Advantages Gained and Lost

So both procedures dealt with in the preceding section avoid Borda's paradox and can thus be considered improvements over the plurality voting. The same is, of course, true of the Borda count which arguably was specifically designed to address Borda's paradox. All three methods (Borda count, plurality runoff and alternative vote), however, correct one major flaw, viz. avoid the choice of an eventual Condorcet loser, but two of them (plurality with runoff and the alternative vote) are accompanied with another flaw that the original culprit, plurality voting, is not plagued with: non-monotonicity. In other words, while in plurality voting the additional support for a winner, *ceteris paribus*, never makes it a non-winner, such an addition may displace winners of the plurality with runoff and alternative vote. Table 4 provides an illustration.

The plurality with runoff procedure first eliminates C whereupon B wins the second counting round. The same choice is made by the alternative vote. Suppose now that the winner B had somewhat more support so that two of the voters with ACB ranking lift B (the winner) at the top of their ranking which then becomes BAC. Then the runoff contestants are B and C with C the winner. The same outcome ensues from the alternative vote procedure. Thus additional support may, indeed, be detrimental for the winning alternative when plurality with runoff or alternative vote are being applied.

In contrast to Nanson's rule, however, the plurality runoff and alternative vote procedures do not necessarily elect the Condorcet winner when one exists. The failure is easy enough to demonstrate. See Table 5 where the Condorcet

Table 5. Plurality with runoff and alternative vote may not elect the Condorcet winner

4 voters	3 voters	2 voters
A	B	C
C	C	A
B	A	B

Table 6. Nanson's rule is non-monotonic

5 voters	4 voters	3 voters	2 voters	2 voters	1 voter
C	B	A	B	A	A
A	D	B	A	C	C
D	C	D	C	B	D
B	A	C	D	D	B

winner C is ranked first by the smallest number of voters and is thus eliminated in the first counting round by both procedures. Hence it is not elected. In fact, with more than three alternatives profiles can be envisaged where the Condorcet winner is ranked first by no voters at all.[5]

As said, Nanson's rule is, by design, Condorcet consistent. So, it is an improvement over plurality with runoff and alternative vote systems in that regard. However, it shares an important drawback with them: it is non-monotonic. See the 17-voter 4-alternative profile of Table 6 where A wins once first D and then both B and C are eliminated. Let now A's support be increased, *ceteris paribus*, so that the 2 voters with BACD ranking lift the winner A on top of their ranking so that their ranking is ABCD. As a result both B and D are eliminated in the first computing round, whereupon C wins.

A property intuitively related to monotonicity is participation. It requires that for any group of identically-minded voters and any profile of preferences, abstaining does not lead to an outcome that is preferable to the one ensuing when the group is voting according to its preferences, *ceteris paribus*. In a seminal paper Moulin established that Condorcet consistency is incompatible with participation if there are more than three alternatives and the number of voters is large enough [12].[6] In other words, all procedures that necessarily elect a Condorcet winner when one exists may encounter profiles of four or more alternatives where a group of voters with identical opinions about alternatives is better off abstaining than voting according to its preferences. Violations of participation are often called no-show paradoxes [6,9].

---

[5] For example, let a three-person four-alternative profile be the following: 1 voter: ABCD, 1 voter: CBDA, 1 voter: DBAC. Here B is the Condorcet winner.

[6] Moulin's lower bound for the number of voters was 25. This bound has more recently been lowered to 12 by Brandt et al. [3]. A stronger variant of Moulin's result has subsequently been proven by Pérez [19].

Table 7. Plurality with runoff and alternative vote fail on participation

7 voters	8 voters	5 voters
A	B	C
B	C	A
C	A	B

Table 8. Nanson's rule may lead to a strong no-show paradox

5 voters	4 voters	3 voters	3 voters	1 voter	5 voters
C	B	A	A	B	C
D	D	C	C	D	A
B	A	D	B	A	D
A	C	B	D	C	B

As Nanson's rule is Condorcet consistent, it is, by Moulin's result, vulnerable to the no-show paradox, but this result as such gives no clues as to whether plurality with runoff or alternative vote are consistent with participation. The following example, however, illustrates the possibility of a no-show paradox in both of the last mentioned procedures in a profile with only three alternatives (Table 7).

Both plurality with runoff and alternative vote result in A after C has been eliminated. This is the worst outcome for the eight voters in the middle. Suppose that four of them had abstained. Then B would have been eliminated, whereupon C would have won. C being preferable to A, we have an instance of the no-show paradox demonstrating that the procedures are vulnerable to it.

A far more dramatic instance of the no-show paradox – called the strong no-show paradox [19] – may occur when Nanson's rule is used. Table 8 gives an illustration. In this 21-voter profile A is the Condorcet winner and is, by Nanson's argument, elected. Suppose now that the five voters with CADB ranking would have abstained. In the ensuing 16-voter profile, there is no Condorcet winner. Instead C emerges as the Nanson winner. Thus, by abstaining the five voters not only improve the outcome (from their point of view), but bring about the victory of their first-ranked alternative.[7]

In terms of the social choice desiderata considered here, the price to be paid for the avoidance of Borda's paradox by plurality with runoff, alternative vote and Nanson's rules seems high: they all suffer from vulnerability to the monotonicity failures both in fixed and variable electorates, while the plurality voting is immune to these anomalies. The same observation can be made about Condorcet consistency, albeit with the significant reservation that Nanson's rule is

---

[7] Table 8 also illustrates the vulnerability to the strong no-show paradox of several other Condorcet consistent procedures: Baldwin, Black, Copeland and Kemeny. For further discussion, see [8].

Condorcet consistent. These conclusions have some bearing on the evaluation of the relatively recent reform proposal concerning the election system of the U.S. House of Representatives.

6 The Reform Proposal of Maskin and Sen

The relatively recent reform proposal by Maskin and Sen is a hybrid voting system intended for the replacement of the current plurality voting system commonly used in the election of the members of the House of Representatives of the U.S. Congress [10].[8] It is pretty similar to the system proposed by Black nearly seventy years earlier [2]. Black's suggestion is simply a combination of Condorcet's and Borda's winning intuitions: given a profile, elect the Condorcet winner if one exists, otherwise elect the Borda winner. In similar vein, Maskin and Sen suggest that the Condorcet winner be elected if one exists, but otherwise the plurality with runoff winner is elected. It is difficult to see how this proposal would improve upon Black's hybrid method. After all, when a Condorcet winner exists in a profile, both Maskin and Sen's procedure (MS, for brevity) and Black's method elect it. The differences can occur only in those profiles where there is no Condorcet winner. In those, MS resorts to plurality with runoff, while Black's method applies the Borda count. The latter is monotonic both in fixed and variable electorates, while the former is non-monotonic in both kinds of electorates. As observed above, both MS and Black's method avoid electing the Condorcet loser.

The primary objective of Black's method and MS is the election of the Condorcet winner. In this sense both procedures are fundamentally majoritarian, i.e. the winner should primarily be determined by pairwise majority comparisons. In the absence of a Condorcet winner, one then resorts to different positional procedures. It is not clear why such a leap from one intuition of winning (binary) to another (positional) is called for. Such a combination of intuitions may lead to quite astonishing occurrences. To wit, the alternative that comes close to being the Condorcet winner may not do well in terms of the plurality with runoff. In fact, even the Condorcet winner may not have sizable support in terms of first ranks of voters. Indeed, as was seen above (footnote 5), it may well be that no voter ranks the Condorcet winner first. Hence if an alternative comes close to being the Condorcet winner (without quite being one), its showing in the plurality with runoff may be the worst.

Overall MS exemplifies the typical tradeoffs involved in rectifying flaws in prevailing voting systems: by correcting one weakness one often ends up with another which may or may not characterize the original system. MS aims at making sure that the Condorcet winners are elected and that the Condorcet losers are not. The original plurality system does not guarantee either of these objectives. Instead, it does well in terms of monotonicity-related desiderata. So by suggesting the replacement of the plurality voting with MS one is in a way revealing a preference for Condorcet criteria over monotonicity-related ones.

---

[8] For a related discussion on the MS procedure, see [18].

Suppose, however, that we could find a procedure that satisfies the Condorcet criteria and does reasonably well in terms of monotonicity. By Moulin's theorem we know the variable electorate variants of monotonicity cannot be satisfied. Yet, there are Condorcet consistent procedures that are monotonic in fixed electorates, e.g. Copeland's method and Kemeny's median, to name two. Intuitively it would make sense to resort to these to guarantee that a uniform standard – the success in pairwise majority comparisons – be applied in determining the winner. This, of course, doesn't do away with the fact that all Condorcet extensions suffer from some form of non-monotonicity in variable electorates [6], but maintains the same winner intuition both in profiles where a Condorcet winner exists and in those where it doesn't.

7 Concluding Remarks

Electoral reforms are often made in order to avoid real or imagined problems in the working of existing procedures. To motivate systemic modifications it is common to concentrate on just one problem at a time rather than to engage in overall evaluation of available procedures in terms of all plausible criteria of performance. The latter, holistic, approach typically reveals theoretical incompatibilities between desiderata. Not all nice properties are achievable under any given procedure in all conceivable profiles. Above we have focused on a few modifications in procedures suggested as responses to observed problems. The tradeoffs involved in these are mainly related to the Condorcet criteria and monotonicity. The problem with these and other similar criteria is their general nature: they deal with all conceivable circumstances. Yet, in practice some circumstances may be excluded or extremely unusual. A plausible way of approaching the electoral reforms would be to take into account any information one might have on such circumstances. For example, are the profiles typically encountered such that a Condorcet winner exists? Are we typically dealing with a small number of alternatives? The vulnerability of various procedures to various anomalies may depend on answers to these kinds of questions [8]. In any event, a holistic multiple criteria evaluations are likely to yield more lasting procedural choices than strategies focusing on one criterion at a time.

References

1. Baldwin, J.M.: The technique of the Nanson preferential majority system. Proc. Roy. Soc. Victoria **39**, 42–52 (1926)
2. Black, D.: On the rationale of group decision-making. J. Polit. Econ. **56**, 23–34 (1948)
3. Brandt, F., Geist, C., Peters, D.: Optimal bounds for the no show paradox via SAT solving. Math. Soc. Sci. **90**, 18–27 (2017)
4. De Grazia, A.: Mathematical derivation of an election system. Isis **44**, 42–51 (1953)
5. Felsenthal, D.S., Machover, M.: After two centuries should Condorcet's voting procedure be implemented? Behav. Sci. **37**, 250–273 (1992)

6. Felsenthal, D.S., Nurmi, H.: Monotonicity Failures Afflicting Procedures for Electing a Single Candidate. Springer, Cham (2017). https://doi.org/10.1007/978-3-319-51061-3
7. Felsenthal, D.S., Nurmi, H.: Monotonicity violations by Borda's elimination and Nanson's rules: a comparison. Group Decis. Negot. **27**(2018), 637–664 (2018)
8. Felsenthal, D.S., Nurmi, H.: Voting Procedures Under a Restricted Domain. An Examination of the (In)Vulnerability of 20 Voting Procedures to Five Main Paradoxes. Springer, Cham (2019). https://doi.org/10.1007/978-3-030-12627-8
9. Fishburn, P.C., Brams, S.J.: Paradoxes of preferential voting. Math. Mag. **56**, 207–214 (1983)
10. Maskin, E., Sen, A.K.: How majority rule might have stopped Donald Trump. The New York Times, 28 April 2016 (2016). https://www.nytimes.com/2016/05/01/opinion/sunday/how-majority-rule-might-have-stopped-donald-trump.html
11. McLean, I., Urken, A.B.: Classics of Social Choice. University of Michigan Press, Ann Arbor (1995)
12. Moulin, H.: Condorcet's principle implies the no-show paradox. J. Econ. Theory **45**, 53–64 (1988)
13. Nanson, E.J.: Methods of elections. Trans. Proc. Roy. Soc. Victoria **19**, 197–240 (1883). McLean, I., Urken, A.B. (eds.) Classics of Social Choice, Chapter 14, pp. 321–359. University of Michigan Press (1995)
14. Niou, E.M.S.: A note on Nanson's rule. Public Choice **54**, 191–193 (1987)
15. Nurmi, H.: On Nanson's method. In: Borg, O., Apunen, O., Hakovirta, H., Paastela, J. (eds.) Democracy in the Modern World. Essays for Tatu Vanhanen, series A, vol. 260, pp. 199–210. Acta Universitatis Tamperensis, Tampere (1989)
16. Nurmi, H.: Voting Paradoxes and How to Deal with Them. Springer, Heidelberg (1999). https://doi.org/10.1007/978-3-662-03782-9
17. Nurmi, H.: Reflections on two old Condorcet extensions. Trans. Comput. Collective Intell. **XXXI**, 9–21 (2018)
18. Nurmi, H.: Voting theory: cui bono? Politeia **91**, 106–121 (2018)
19. Pérez, J.: The strong no show paradoxes are a common flaw in Condorcet voting correspondences. Soc. Choice Welfare **18**, 601–616 (2001)
20. Saari, D.G.: Capturing 'The Will of the People'. Ethics **113**, 333–349 (2003)

Repeated Trust Game – Statistical Results Concerning Time of Reaction

Anna Motylska-Kuźma$^{(\boxtimes)}$ ⓘ, Jacek Mercik ⓘ,
and Aleksander Buczek ⓘ

WSB University in Wroclaw, Wroclaw, Poland
{anna.motylska-kuzma, jacek.mercik}@wsb.wroclaw.pl,
Aleksander.buczek@gmail.com

Abstract. The paper presents basic results regarding probability distributions together with the parameters related to the decision-making time in the repeated trust game. The results obtained are of a general nature, related to the waiting time for a reaction in computer-aided systems, as well as a special one related to the characteristics of the decision-makers participating in the experiment.

1 Introduction

Time is the dimension in which all living organisms adjust their environment [2]. Some of this adjustments take generations and even millenniums, while others only hours, minutes or fractions of seconds. The current work takes into account one of the most important mechanisms of the fast adjustments, namely decision making. The process is connected with the ability to react to information and to take one of a few different action alternatives. Thus the decision process consumes time for processing information. Some decision are very fast, and the decisions in many case emerge with lightning speed. Such decision are habitual or intuitive non-analytic decisions (also called heuristic decisions), which are not based on extensive information-processing [23]. The more a decision process is analytic and algorithmic, the more time is needed for its utilization. For example, when decision expects to be calculated expected value of many alternatives and choose only one with the highest value, it is time consuming to calculate all the alternatives values and then choice the best one. It seems that many day-to-day decisions fall in one of two categories: some decisions take a very short duration and in fact seem automatic (e.g. driving, typing, deciding what to eat, etc.), and others take a very long time and seem very laborious (e.g. where to go on the vacation, what stocks to buy and sell, what investment to choose, etc.). The duration needed to make the decision is only one of possible relationship between time and decision making process. Taking time into account within the decision process can be also in term of considering the optimal time to make a decision or the changes in the decision structure as a function of time [2]. So called dynamic decisions are very common in day-to-day life.

The dynamic decision making process (DDM) is an answer to dynamic changing environment, which vary in their inclusion of delayed feedback, interlinked actions and their effects over time, and time dependence, where the value of actions is determined

© Springer-Verlag GmbH Germany, part of Springer Nature 2019
N. T. Nguyen et al. (Eds.): TCCI XXXIV, LNCS 11890, pp. 74–89, 2019.
https://doi.org/10.1007/978-3-662-60555-4_6

by when an action is taken. The accumulation of these characteristics makes an environment dynamic and complex to different degrees [10]. Although not all the DDM tasks involve all this characteristics, every dynamic decision task must involve a series of choices taken over time to achieve some overall goal. Conceptually, dynamic decision making is a closer learning loop in which decision are informed by the results of previous choices and their outcomes [9, 11].

Current research about DDM processes are focused of the effect of knowledge, experience and intuition in decision making. They investigate the effects of context and properties of the decision environment as well the collective behaviours rather than individual behaviour alone [10]. The results show at least three factors that influence the human exploration process. First, it has been found that people engage in very limited exploration before making a choice (e.g. [10, 13, 14]), but people search longer when they encounter a prospect of losses and when they experience variant relative to consistent environments [16, 20]. Second, people often fail to maximize payoffs, and rather, people often much their response probabilities to the payoff probabilities [8, 17, 24]. Third, people learn to adapt to changing outcomes and probability distributions, but adaptation can be slow and it depends on cognitive parameters of the information experienced, such as the recency and primacy of the relative outcomes from different alternatives [6, 18, 20, 22].

In our research we used the repeated trust game to analyse the duration of decision making in dynamic environment and its connections with chosen parameters of experiments and the characteristics of the players. We want to determine if the decision-making time is related to the nature of the game (infinitely repeated game vs finitely repeated game) as well as determine the general characteristics of decision-making time in economic experiments.

According to Ariely and Zakay [2] if we assume the perfect knowledge and memory, we could rationally expect that time would not have any effect on evaluations and decisions. Nevertheless, from the human point of view it does not exist such thing like perfect knowledge and memory. Thus it is interesting to know, what is the relationship between the time and the decision and between the duration of decision process and the characteristics of the players.

2 Repeated Trust Game - The Experimental Design

Our experiment (details may be found in [21]) is based on the trust game according to Berg et al. [3], but the decisions are made sequentially not simultaneously and the endowment is not equal. Player 1 is given some amount of money (a), which he can transfer (x1) in some part or whole to Player 2. The value of the money he transfers will be multiplied by multiplier (m). Player 2 then decides how much money (x2) to transfer back to Player 1. Thus the payoff resulting from the decisions are given by

$$v1(x1, x2) = a - x1 + x2 \tag{1}$$

$$v2(x1, x2) = m\,x1 - x2 \tag{2}$$

where $x1 \epsilon$ {0, 1, 2, ..., a}, $x2 \epsilon$ {0, 1, 2, ..., m·x1}, a > 0, m > 0.

Actual behaviour of the players in trust games results from the level of trust and level of aversion to inequality. Studies have shown that the amount of transferred money by Player 1 is associated with age and knowledge regarding the responder [1, 12, 19, 25]. In our article, we try to answer whether also the time of decision depends on these factors.

It has to be clearly emphasized that in contrast to the previous research devoted to such games in our schematic, it is possible (and intentionally it was done) to repeat the matches of the same player pairs with both the same and changed values of general game parameters, i.e. basic amount of money (a), as well as multiplier (m). Note also that it is advisable to change roles in individual pairs, i.e. to be the first or the second player, so that it is possible to examine players' behavior in the awareness of the role played in a given game not only at a given moment, but also taking into account possible roles in the future. In the classic approach to the trust game, there was no such possibility.

For the implementation of experiments we have created the appropriate software to support and automate the course of the game. At the beginning, the participants of the game made registration with the provision of personal data: name, surname, education, age, working status (e.g. employee, student, etc.), the employer's data (e.g. sector, size) and the position (i.e. manager or not). Each participant had a unique login and password to enable bidding and viewing the results of previous games. Then, a random selection of player pairs (j) and setting of game parameters was made, i.e. basic amount (a_j) and multiplier (m_j). The course of the game was as follows:

- Player 1 was informed by the system about the conditions of the game. He/she knew the basic amount (a_j), multiplier (m_j), the maximum number of rounds (k). He/she had also the information that the roles will be changed and if the player is first, in the next round he/she could be the second. According the data, he/she could decide to transfer the amount $(x1_j)$ to the Player 2. For a decision the Player 1 had one week, which means that he/she could seriously think over his/her response,
- Player 2 was informed by the system about the decision of Player 1. He/she knew the basic amount (a_j), multiplier (m_j), the maximum number of rounds (k). He/she had also the information that the roles will be changed and if the player is second, in the next round he/she could be the first. Player 2 had also one week to make a decision and transfer of the selected amount $(x2_j)$ to Player 1 from the same pair,
- after the end of a given game individual results were saved on the account of each participant of the game. Each of them could at any time view their results and analyze their previous behavior.

In the Figs. 1, 2, 3, and 4 are shown the basic characteristics of the total number of participants. We can see that most of them have a high school degree, are less than 50 years old and are employed. From the point of view of gender, there are more women than men.

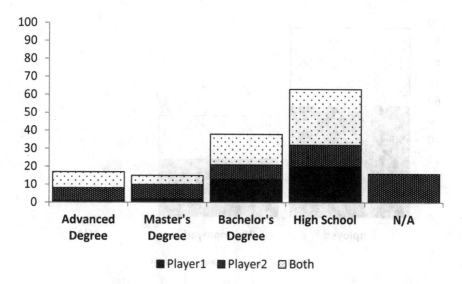

Fig. 1. Participant education level by player role

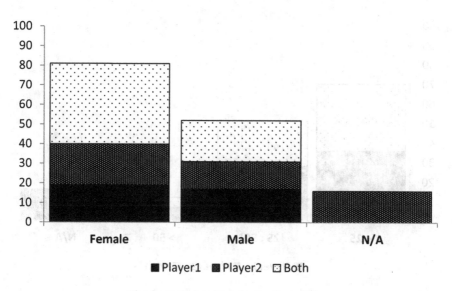

Fig. 2. Participant gender by player role

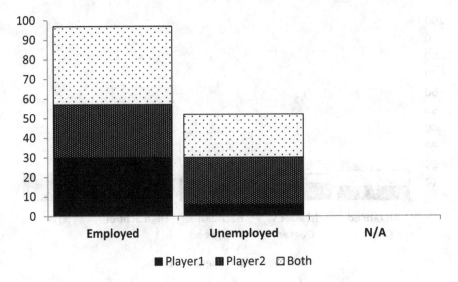

Fig. 3. Participant employment status by player role

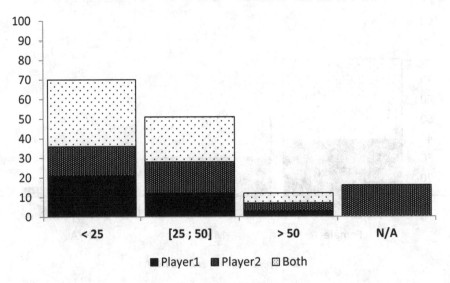

Fig. 4. Participant age by player role

Figure 5 (below) shows the dynamics of the game due to the passage of time. From the beginning, the players knew that there would be more than one round in a given game, however until the third round they did not know that the whole game will end after four rounds. Therefore, it can be assumed that the first two rounds were played as infinitely repeated game and the last two finitely repeated game. One of our goals is to determine whether this was reflected in the time players spent on making decisions.

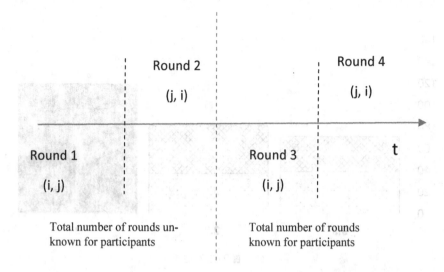

Fig. 5. Description of the course of experiments. For given parameters (basic amount and multiplier) 4 rounds were executed however the total number of rounds was unknown to the players for two first rounds.

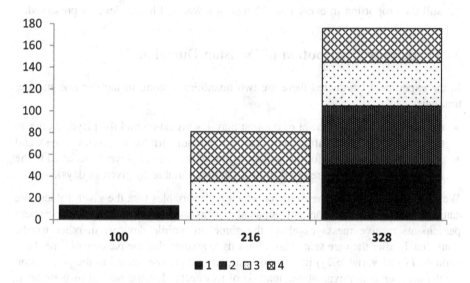

Fig. 6. Basic amount dimension by the round number

Figure 6 shows the dimension of the basic amount by the round number. We can observe that only in the first round the basic amount is the lowest (100). In further rounds it increases, however there are proportional number of observation with lower and higher level of basic amount within every round. In the Fig. 7 we can observe opposite effect in the dimension of the multipier. Only in the first round the multiplier

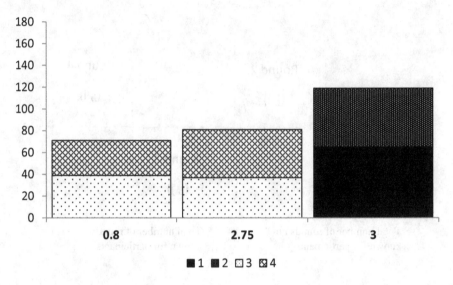

Fig. 7. Multiplier dimension by the round number

has the highest level. The further rounds characterise with the lower value of multiplier, but still the proportion in every round between lower and higher level is preserved.

3 Probability Distribution of Decision Duration

In the experiment conducted there are two quantities relating to the decision-making time, viz:

- the time between the receipt of the first player's invitation and the player's offer to the other player (hereinafter referred to as a random variable T_1 given in days), and
- analogous waiting time for the analogue offer of the second player transferred to the first player (hereinafter referred to as a random variable T_2 given in days).

We do not know the exact reasons for which both variables take the given values. We can only assume that these are not technical reasons, as the vast majority of experiment participants receive messages about the game on mobile devices, in other words, immediately after they are sent. This allows us to assume that the passage of time (both variable T_1 and variable T_2) in a significant part is therefore related to the preparation for the decision (offer evaluation, analysis of its effects, development of own strategy, etc.) and not the technical aspect of the process of communicating with the system itself, which is simple and does not require significant time to carry it out. In any case, the time to make a decision is limited to one week (Tables 1 and 2).

Analysing both random variables firstly we will examine the type and parameters of the probability distribution describing them. In the next section we will try to answer the question whether the observed time values can be linked to specific parameters observed in the game. Thus, we believe that the type of probability distribution of a

Table 1. Summary Statistics for the variable duration of the first decision (271 valid observations)

Mean	Median	Minimum	Maximum
1.53937	0.966586	0.406424	6.84655
Std. Dev.	C.V.	Skewness	Ex. kurtosis
1.20858	0.785112	1.95383	4.38202
5% Perc.	95% Perc.	IQ range	Missing obs.
0.470100	3.85092	1.11245	0

Source: own calculations (The calculations were made with the program GRETL https://sourceforge.net/projects/gretl/ (available 10.01.2019).)

Table 2. Summary Statistics for the variable duration of the second decision (79 valid observations)

Mean	Median	Minimum	Maximum
0.70664	0.21919	0.00064815	5.6763
Std. Dev.	C.V.	Skewness	Ex. kurtosis
1.0136	1.4344	2.3000	6.7190
5% Perc.	95% Perc.	IQ range	Missing obs.
0.0017593	2.7241	1.0336	0

Source: own calculations (The calculations were made with the program GRETL https://sourceforge.net/projects/gretl/ (available 10.01.2019).)

time variable is a more general feature and its knowledge may also be useful in describing other situations related to computer experiments, in which participants are expected to respond appropriately, closing between fixed time moment.

Analysing the time devoted to the implementation of the decision-making process, we notice some of its features that are related to the essence of time as such: it accepts only positive values and its termination (even if all such processes start at the same moment), in most cases is right-side variable[1] and in our case, limited (truncated) to a given size. Thus, it should be assumed that the random variables similar to variables T_1 and T_2 should belong to the class of continuous probability distributions and probably be similar to the four-parameter Beta probability distribution (e.g. [15]):

$$f(x) = \frac{ax^{ap-1}}{b^{ap}B(p,q)(1 + (x/b)^a)^{p+q}}$$

(3)

---

[1] Similar relationships were observed e.g. in the study Personal Income Distribution ([7], in actuarial losses estimation [15] or examining the duration of a company merger [4].

where $x > 0$, $a > 0$, $b > 0$, $p > 0$, $q > 0$ and $B(p, q) = \int_0^1 t^{p-1}(1-t)^{q-1}dt$ denotes Beta function.

Therefore, we examined the results obtained referring to the decision-making time regarding compliance with the following probability distributions being Beta distribution mutations: Burr type XII, Gamma, Generalized Pareto Distribution, Log-Cauchy, Log-normal, Weibull, Dagum (Burr type III), Gauss, Pareto, Wald and Exponential[2]. The results obtained indicate that:

- for a random variable T_1 the best fit occurred for the Dagum distribution (according to all the most commonly used matching quality measures, i.e. Akaike, Kolmogorov-Smirnov, Cramer-von Mises, Anderson-Darling)

$$f(x) = \frac{\alpha\beta x^{\alpha\beta-1}}{(1+x^\alpha)^{\beta+1}} \tag{4}$$

where $x > 0$, $\alpha > 0$, $\beta > 0$,
- for a random variable T_2 the best fit occurred for the Gamma distribution (according to the criterion Akaike, statistics Anderson-Darling and Cramer-von-Mises) or Log-normal (Kołmogorow-Smirnov statistic)

$$f(x) = \begin{cases} \frac{\lambda^p}{\Gamma(p)} x^{p-1} e^{-\lambda x} & \text{for } x \geq 0 \\ 0 & \text{for } x < 0 \end{cases} \tag{5}$$

for $p, \lambda > 0$ and $\Gamma(p)$ denotes gamma function (Figs. 8, 9 and 10).

As we have already mentioned, decisions are made within a given one-week interval from the moment of the call to take them. This raises the suspicion that the data we have is censored data (right-truncated)[3]. Particularly, it seems to be the case for a variable T_1, referring to a situation in which there is a significant number of non-decision cases (despite the previous declaration regarding participation in the experiment): 79 decisions in the second step compared to 271 decisions in the first step. Analyzing the variable T_2 assuming that this is a random truncated variable, we find that the best fit is characterized by the Burr XII distribution[4] (statistics of Kolmogorov-Smirnov, Cramer von Mises and the measure of a two-step logarithm of likelihood).

$$f(x) = \frac{\alpha\beta x^{\alpha-1}}{(1+x^\alpha)^{\beta+1}} \tag{6}$$

where $x > 0$, $\alpha > 0$, $\beta > 0$.

---

[2] The description of the distributions can be found, for example, in the SAS documentation: http://support.sas.com/documentation/cdl/en/etsug/63939/HTML/default/viewer.htm#etsug_severity_sect017.htm (14.01.2019).

[3] As it is defined in the SAS documentation: http://support.sas.com/documentation/cdl/en/etsug/63939/HTML/default/viewer.htm#etsug_severity_sect018.htm (19.01.2019).

[4] This distribution is known as the Burr distribution [5].

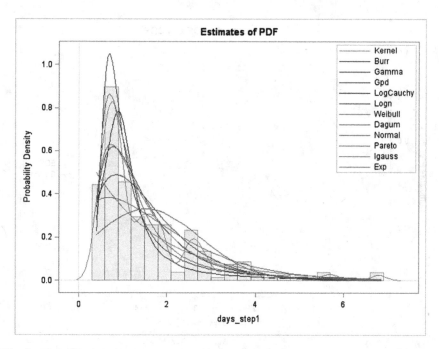

Fig. 8. Probability distribution functions for random variable T_1 (source: own calculations).

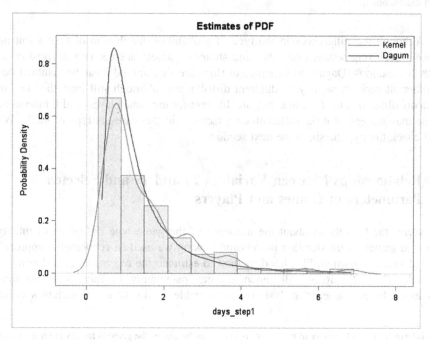

Fig. 9. Probability distribution function for random variable T_1 Dagum distribution (source: own calculations).

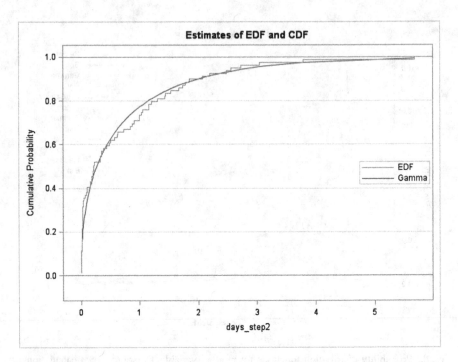

Fig. 10. Cumulative distribution function for random variable T_2 - Gamma distribution (source: own calculations).

Although the differences in the type of probability distribution of time spent on making decisions between one-shot and strategic games[5] are not very surprising, as both distributions (Dagum and Gamma or Burr) are similar (and it can be assumed that in other attempts these may be different distributions, although still from the class of Beta distributions), the fact, that they are different for the same players and for the same game may suggest that the differentiating factor is in this case the type of game. We will check the hypothesis in the next section.

4   Relationships Between Variables T_1 and T_2 and Selected Parameters of Games and Players

To verify the hypothesis about the influence of the game type (infinitely vs finitely repeated game) on the duration of decision-making we used an econometric approach related to the maximum likelihood method to estimate the regression of a dependent variable (T_1 or T_2) against all known players' parameters. In Table 3 are the final results of the econometric analysis for the variable T_1 (in the set of variables called

---

[5] If information about number of rounds (k) is known to the player the game is finitely repeated game: the player should calculate his/her payoff anticipating future results. Otherwise the game is infinitely repeated game.

days_step1); in Table 4 similar results for the variable T_2 (in the set of variables called days_step2). As could be seen, both econometric models differ significantly.

Table 3. Econometric model of duration for the first decision (T_1).

	Coefficient	Std. Error	t-ratio	p-value	
const	0.207748	0.206718	1.005	0.3158	
Basicamount	−0.00148460	0.000595374	−2.494	0.0133	**
First player's age	0.0185766	0.00468844	3.962	<0.0001	***
First player's education	−0.174667	0.0509616	−3.427	0.0007	***

Mean dependent var	0.194396	S.D. dependent var	0.663237
Sum squared resid	109.4013	S.E. of regression	0.640111
R-squared	0.078869	Adjusted R-squared	0.068519
F(3, 267)	7.620371	P-value(F)	0.000066
Log-likelihood	−261.6208	Akaike criterion	531.2416
Schwarz criterion	545.6501	Hannan-Quinn	537.0268

Table 4. Econometric model of duration for the second decision (T_2).

	Coefficient	Std. Error	t-ratio	p-value	
const	2.08971	0.715969	2.919	0.0046	***
Multiplier	−0.249526	0.139727	−1.786	0.0781	*

Mean dependent var	0.706635	S.D. dependent var	1.01359
Sum squared resid	75.85462	S.E. of regression	0.99904
R-squared	0.053422	Adjusted R-squared	0.02851
F(2, 76)	2.144603	P-value(F)	0.12414
Log-likelihood	−110.4913	Akaike criterion	226.982
Schwarz criterion	234.0909	Hannan-Quinn	229.830

In the model for variable T_1, dependent variables parameters related to basic amount are played: age and education of players (with the first move). In this respect, the results confirm the findings obtained earlier in the works [1, 12, 19, 26], although the earlier ones are referred only to payoffs. Therefore, also the duration of decision making depends on the parameters mentioned (age and education).

We get surprising and diametrically different results for the variable T_2: duration of decision-making is not dependent on factors related to players. There is only a dependence on the multiplier and even this is of relatively low significance. The parameter, which is the multiplier, however, is a parameter related to the whole game. What's more, research shows that the higher the multiplier value, the shorter the decision-making time. Since the multiplier has a decisive impact on the payoff, we would expect a reverse dependency: a larger amount of payment (as being more "serious") should require a longer time to think about the decision. However, this is not the case in our observations and requires further research.

Analysing the average decision duration of the same players making decisions as the first players (round 1 – unknown k, and round 3 – known k), we also find that the hypothesis about the equality of the mean duration for unknown and known number of rounds (k) must be rejected (p-value is 4.8E−06). Thus, we find that knowledge of the number of games is more important for the time spent on making the player's decision in the first position.

A similar hypothesis about the equality of time devoted to making decisions for players with some experience (they already had participation in the competition but in the second position) in the first position (round 2 and 4) is not so clear (p-value is 0.03634 for single-side test and 0.07268 for a two-side test), although the hypothesis of equality of averages can still be rejected.

Similar results are obtained for players who play as second. Summing up, therefore, the knowledge of the number of games and (less well) experience gained earlier influences the duration of making decisions. In both cases the decision's duration is extended.

5   Conclusions and Future Research

Analysis of the duration of the decision-making in the Repeated Trust Game leads to the following conclusions:

- The duration of the decision-making has a Burr type probability distribution (III or XII type). As can be seen, the use of symmetrical Beta probability distributions (for example such as estimating the duration of activities in the PERT method) is not justified. As it seems, this conclusion should also be true for all other times related to making decisions in computer aided systems and it means that estimations usually given by experts may be imprecise. This of course requires further research.
- Knowledge about the number of games extends the duration of decision-making. According to intuition it should be assumed that one-shot games require shorter decision-making times than strategic games, and that this observation is general in nature and applies not only to the Repeated Trust Game.
- In our experiment of Repeated Trust Game the multiplier has a decisive impact on the payoff but we obtain a reverse dependency: a larger amount of payment (as being more "serious") should require a longer time to think about the decision. However, this is not the case in our observations and requires further research.
- The size of the payoff reduces the duration of decision making if the number of games is known. Is it something like impatience associated with the winnings or "carelessness" resulting from the fact that any errors can still be corrected?

Limitations

Our experiment and the results have some limitations, e.g.:

- In our experiment the players were informed about the points which are given to them or which they decide to give to the counterplayer. In the original trust game it is used the amount of money. Therefore, from the psychological point of view, this is not the same and drive not the same behavior. Furthermore, the time spend for the decision could be not equal to this devoted on decision concerning the money. In the future experiments it will be interesting to check the differences between the time spend for decision making if the player take into account the points versus money.

- Each pair plays the trust game only four times (twice in both roles). It seems that if we divide the game into infinitely/finitely repeated game with and without history we have only one observation for one player in each role (player 1 in infinitely repeated game without history, player 2 in infinitely repeated game with some history cause from changing the role, player 1 in finitely repeated game, player 2 in finitely repeated game). Thus, we cannot conclude about the strategic thinking of the players (e.g. learning process, seizing the opportunities, etc.), what could be interesting to analyse.

- It should be noticed that the decision in the final round could be dependent from two factors: (a) analysis of the history of the game and (b) the fact that it is the last round. In consequence, the fourth round could be interpreted as a one-shot game with previous history, not a repeated game. If it were, we could compare the time spend for decision making in one-shot game and the repeated game. However, it is impossible in our case because of too few rounds played.

- The time considered in our experiment include the time that passes before the decision is made and the decision itself. This could be misleading, because some decision could be made instinctively and some are made after reflection. Separating this two phase of decision process and concerning the time spend for each of them could give us the information about the relations between the time required to make a decision and the decision itself. Thus we can analyse the "instinctive" and "reflection" decision. It would be interesting to know what transfer by Player 1 lead to "instinctive" decisions and whether "instinctive" decisions differ from the "reflective" decisions.

It should be emphasized that the repetition of games resulting from the fact that the experiment is supported by a computer system ensures the participation of the same pair of players in changing circumstances. This allows to verify some of the already mentioned assumptions relating to the characteristics of the Repeated Trust Game. We intend to carry out such verification in the future. Another aspect of the repetitive research, which we will deal with in the near future is the appropriate profiling of the participants of the game so that it is possible to determine the relationship between the characteristics of the players and the results obtained.

References

1. Akai, K., Netzer, R.J.: Trust and reciprocity among international groups: experimental evidence from Austria and Japan. J. Socio-Econ. **41**, 266–276 (2012)
2. Ariely, D., Zakay, D.: A timely account of the role of duration in decision making. Acta Physiol. **108**, 187–207 (2001)
3. Berg, J., Dickhaut, J., McCabe, K.: Trust, reciprocity and social history. Games Econ. Behav. **10**, 122–142 (1995)
4. Buczek, A.: The time to completion of a legal merger: general concepts, statistical analysis and the case of Poland. Oper. Res. Decisions **26**(1), 19–44 (2016)
5. Burr, I.W.: Cumulative frequency functions. Ann. Math. Stat. **13**(2), 215–232 (1942)
6. Cheyette, S., Konstantinidis, E., Harman, J.L., Gonzalez, C.: Choice adaptation to increasing and decreasing event probabilities. Paper presented at the 38th Annual Meeting of the Cognitive Science Society (CogSci 2016), Philadelphia, PA (2016)
7. Dagum, C.: A new model of personal income distribution: specification and estimation. Economie Appliquée **30**, 413–437 (1977)
8. Erev, I., Barron, G.: On adaptation, maximization, and reinforcement learning among cognitive strategies. Psychol. Rev. **112**, 912–931 (2005)
9. Gonzalez, C., Meyer, J.: Integrating trends in decision making research. J. Cogn. Eng. Decis. Making. Advance online publication (2016). https://doi.org/10.1177/1555343416655256
10. Gonzalez, C., Fakhari, P., Busemeyer, J.: Dynamic decision making: learning processes and new research directions. Hum. Factors **59**(5), 713–721 (2017). https://doi.org/10.1177/0018720817710347
11. Gonzalez, C., Lerch, J.F., Lebiere, C.: Instance-based learning in dynamic decision making. Cogn. Sci. **27**, 591–635 (2003)
12. Guillen, P., Ji, D.: Trust, discrimination and acculturation. Experimental evidence on Asian international and Australian domestic university students. J. Socio-Econ. **40**, 594–608 (2001)
13. Hertwig, R., Pleskac, T.J.: The game of life: how small samples render choice simpler. In: Chater, N., Oaksford, M. (eds.) The Probabilistic Mind: Prospects for Bayesian Cognitive Science, pp. 209–235. Oxford University Press, Oxford (2008)
14. Hertwig, R., Pleskac, T.J.: Decisions from experience: why small samples? Cognition **115**, 225–237 (2010)
15. Kleiber, C., Kotz, S.: Statistical Size Distributions in Economics and Actuarial Sciences. Wiley, New Jersey (2003)
16. Lejarraga, T., Dutt, V., Gonzalez, C.: Instance-based learning: a general model of repeated binary choice. J. Behav. Decis. Making **25**, 143–153 (2012)
17. Lejarraga, T., Hertwig, R., Gonzalez, C.: How choice ecology influences search in decisions from experience. Cognition **124**, 334–342 (2012)
18. Lejarraga, T., Lejarraga, J., Gonzalez, C.: Decisions from experience: how groups and individuals adapt to change. Mem. Cogn. **42**, 1384–1397 (2014)
19. Markowska–Przybyła, U., Ramsey, D.: A Game Theoretical Study of Generalised Trust and Reciprocation in Poland. I. Theory and Experimental Design. Operations Research and Decisions, No. 3 (2014)
20. Mehlhorn, K., Ben-Asher, N., Dutt, V., Gonzalez, C.: Observed variability and values matter: towards a better understanding of information search and decisions from experience. J. Behav. Decis. Making **27**, 328–339 (2014)
21. Motylska-Kuzma, A., Mercik, J., Sus, A.: Repeatable trust game – preliminary experimental results. In: Nguyen, N.T., et al. (eds.) TCCI 2019. LNCS, vol. 11890. Springer, Cham (2019)

22. Rakow, T., Miler, K.: Doomed to repeat the successes of the past: history is best forgotten for repeated choices with non-stationary payoffs. Mem. Cogn. **37**, 985–1000 (2009)

23. Russo, J.E., Schoemaker, P.J.: Decision Traps: Ten Barriers to Brilliant Decision-Making and How to Overcome Them. Doubleday/Currency, New York (1989)

24. Shanks, D.R., Tunney, R.J., McCarthy, J.D.: A re-examination of probability matching and rational choice. J. Behav. Decis. Making **15**, 233–250 (2002)

25. Singh, S.K., Maddala, G.S.: A function for size distribution of incomes. Econometrica **44**(5), 963–970 (1976)

26. Slonim, R., Garbarino, E.: Increases in trust and altruism from partner selection: experimental evidence. Exp. Econ. **11**, 134–153 (2008)

Labeled Network Allocation Problems. An Application to Transport Systems

Encarnación Algaba[1]([⊠]), Vito Fragnelli[2], Natividad Llorca[3],
and Joaquin Sánchez-Soriano[3]

[1] Department of Applied Mathematics II and IMUS,
University of Seville, Seville, Spain
ealgaba@us.es

[2] Department of Science and Innovative Technologies,
University of Eastern Piedmont, Viale Teresa Michel 11, 15121 Alessandria, Italy
vito.fragnelli@uniupo.it

[3] Research Institute CIO and Department of Statistics,
Mathematics and Computer Science,
University Miguel Hernández of Elche, Elche, Spain
{nllorca,joaquin}@umh.es

Abstract. We deal with networks in which there are more than one arc connecting two nodes. These multiple arcs connecting two nodes are labeled in order to differentiate each other. Likewise, there is traffic or flow among the nodes of the network. The links can have different meanings as such roads, wire connections or social relationships; and the traffic can be for example passengers, information or commodities. When we consider that labels of a network are controlled or owned by different agents then we can analyze how the worth (cost, profit, revenues, power...) associated with the network can be allocated to the agents. The Shapley quota allocation mechanism is proposed and characterized by using reasonable properties. Finally, in order to illustrate the advantages of this approach and the Shapley quota allocation mechanism, an application to the case of the Metropolitan Consortium of Seville is outlined.

Keywords: Allocation mechanisms · Networks · Shapley quota allocation mechanism

1 Introduction and Literature Review

Networks are very often used to graphically represent many different situations from social relationships to real physical problems such as road maps. For example, in Operations Research and Management Science, due to the possibilities offered by such a representation, it is commonplace. We draw attention to the volumes by Ball et al. (1995a and 1995b) for a very informative survey. Networks also play an important role to analyze social and economic problems (see, for example, Megiddo 1978; Sharkey 1995; Slikker and van den Nouweland 2001;

© Springer-Verlag GmbH Germany, part of Springer Nature 2019
N. T. Nguyen et al. (Eds.): TCCI XXXIV, LNCS 11890, pp. 90–108, 2019.
https://doi.org/10.1007/978-3-662-60555-4_7

Jackson and Zenou 2014; Algaba et al. 2017, 2018). Networks are also used to model the interaction between parts of information or computer systems (see Tardos (2004) and references herein). Therefore, network models are interesting enough to go further in their analysis.

In this paper, we consider networks in which there is one perfectly divisible unit of flow or traffic between different nodes of the network. Somehow, we have multi-flow networks because the flow between each pair of nodes can be considered different from others. Likewise, multiple arcs connecting two nodes are allowed, for this reason they are labeled in order to distinguish each other. The part of the unit of flow between two nodes can go throughout different routes and this is known due to different reasons, for example capacity conditions. In the network there are agents who control different sets of labels. Thus, the set of arcs of the network is partitioned among the agents. This network model is more general than the one introduced by Algaba et al. (2019a) to study the profit allocation problem in horizontal cooperation in public transport systems. We call *labeled networks* the situations described above and *labeled network allocation problem* the problem of allocating the worth associated with the network among the agents controlling its different components.

Moreover, allocation network problems start from an existing network, and often deal with the problem of allocating profits and/or costs for building and/or maintaining the network among the users. For example, Granot and Hojati (1990) study how to allocate the cost of constructing a communication network. They consider two possible situations and for both determine the nucleolus and the Shapley value. Tijs et al. (2006) study the Bird core correspondence for minimum cost spanning tree games. Koster et al. (2001) study the core of standard fixed tree games and prove that the core of these games coincides with the set of all weighted constrained egalitarian solutions. Bjorndal et al. (2004) study standard fixed tree games, for which each vertex unequal to the root is inhabited by exactly one player, and give an alternative proof of that their cores equal their corresponding sets of weighted Shapley values. Gupta et al. (2004) study how to define good cost-sharing mechanisms for single-source network design problems. Maschler et al. (2010) introduce a new algorithm to compute the nucleolus of standard tree games. Bergantiños et al. (2014) introduce an allocation rule to divide the cost of a network which connects the agents to a service provided from a source. Roughgarden and Schrijvers (2014) study network cost-sharing games in which the cost of each edge is shared using the Shapley value. They then study the equilibria of the associated non cooperative games. In all the previous papers, how to distribute the cost of building or maintaining a network that connects the agents to a source that provides a useful service is studied, while in this paper we study how to distribute the known flow that circulates through a network between the agents who control that network. On the other hand, there is a number of papers that study from a game theoretical point of view flow problems in networks. Kalai and Zemel (1982), Curiel et al. (1989) and Reijnierse et al. (1996) study the nonemptiness of the core of different simple flow games. Derks and Tijs (1985, 1986) study the case of multi-commodity flow situations and also study the nonemptiness of the core of associated games. In all these papers, how to distribute the maximum flow that can be obtained from cooperation between the agents that control the network is studied, while in this paper we study how to distribute the flow (or the profit/cost associated with this flow), which has effectively occurred, among the agents that control the network. In addition, we focus on the allocation problem rather than studying the game associated with this problem.

Finally, in order to illustrate how our network model can be applied to a real-life situation, we consider the case of the Metropolitan Consortium of Seville. In particular, a reduced and stylized situation from the real transport system is simulated. In this case, when the members of the consortium cooperate, 405 feasible routes connecting different points of the transport network in the city center are determined, but only 92 of these feasible routes are operated by a single company. Therefore, the advantage of cooperation is clear. For this problem, we propose the Shapley quota allocation which takes into consideration not only the number of routes, but also the traffic flow. Furthermore, this solution is compared with a proportional distribution based on the number of routes in which each company is involved. Some interesting examples which relate to this situation are the following. Fragnelli et al. (2000b) study how to share the profit of a shortest path situation and Sánchez-Soriano (2003 and 2006) proposes two solutions for the profit allocation problem arising from the classic transport problem, based on pairwise distributions. We can also find real-life applications to transport situations in which there is an underlying network. For example, Fragnelli et al. (2000a) and Norde et al. (2002) study how to allocate the cost of a railway line used by different trains, each of which has different needs and requirements; and Sánchez-Soriano et al. (2002) study how to share the cost of a public transport system for students in the area of Alicante.

The rest of the paper is organized as follows. In Sect. 2, we introduce the network allocation problem which we analyze in this paper. In Sect. 3, the Shapley quota allocation mechanism for labeled network problems is characterized by using reasonable properties related to the context of networks. In Sect. 4, the labeled network allocation problem is applied to the Metropolitan Consortium of Seville and the Shapley quota allocation is computed and commented. Section 5 concludes.

2   The Labeled Network Allocation Problem

We consider networks in which there is one perfectly divisible unit of flow or traffic between different nodes of the network. Somehow we have multi-flow networks because the flow between each pair of nodes can be considered different from others. Likewise, multiple arcs connecting two nodes are allowed, for this reason they are labeled in order to distinguish each other. The part of the unit of flow between two nodes can go throughout different routes and this is known due to different reasons, for example capacity conditions. In the network there are agents who control different sets of labels. Thus, the set of arcs of the network is partitioned among the agents. The worth obtained by a subset of agents is the part of the unit of flow that they can obtain by using only their arcs.

Formally, a *labeled graph* is described by the 3-tuple $\mathcal{G} = (V, L, A)$, where V is a finite set of nodes; L is a finite set of labels; and $A \subset V \times V \times L$ is a finite set of labeled directed arcs connecting nodes of V, such that $(i, j, l) \in A$ means that i is the initial node, j is the end node and l is the label. Moreover, we assume that $(i, i, l) \notin A, \forall\, i \in V, \forall\, l \in L$, i.e., loops are not allowed.

A *labeled route* connecting two nodes $i, j \in V$ in a labeled graph (V, L, A) is a sequence of labeled arcs $\{(i, i_1, l_1), (i_1, i_2, l_2), ..., (i_{k-1}, j, l_k)\} \subseteq A$.

Let $\mathcal{R}$ be a set of feasible labeled routes connecting two nodes of V. $\mathcal{R}$ could be a proper subset of the set of all possible labeled routes connecting two nodes of V, $\mathcal{R}(A)$. In this situation, some labeled routes would have been discarded because they are useless or impossible. Let f be a function describing how a (perfectly divisible) unit of flow is distributed throughout all labeled routes, i.e., $f : \mathcal{R}(A) \to [0, 1]$ such that

- $f(r) = 0$, if $r \notin \mathcal{R}$, and $f(r) \geq 0$, if $r \in \mathcal{R}$.
- $\sum_{r \in \mathcal{R}} f(r) = 1$.

Therefore, we assume that the distribution of the unit of flow throughout the graph is perfectly determined. This could occur due to different reasons, for example, capacity constraints, ex-post observation of the traffic, preferences of individuals using the network, a centralized management of the network controlling the traffic throughout the graph, an exogenous condition, etc. Likewise, this function f can be derived from an origin-destination (OD) matrix and it can measure the probability of a particular labeled route to be used in the network.

A labeled network arises when we consider a labeled graph and the flow throughout it. Therefore, a *labeled network* is described by the 3-tuple $\mathcal{N} = (\mathcal{G}, \mathcal{R}, f)$, where $\mathcal{G} = (V, L, A)$ is a labeled graph, $\mathcal{R}$ is a set of feasible labeled routes and f is a distribution of one unit of flow among all feasible labeled routes.

We now provide a simple example to illustrate the different elements of a labeled network.

Example 1. *Consider a simple network as depicted in Fig. 1 with 5 nodes, three labels R (continuous line), B (dotted line) and G (dashed line) and the arcs always go from i to j such that $i < j$.*

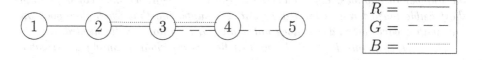

Fig. 1. A simple network

The possible origin-destination (OD) pairs are 1-2, 1-3, 1-4, 1-5, 2-3, 2-4, 2-5, 3-4, 3-5 and 4-5.

We can consider the following set of non-exhaustive labeled routes:

OD	$Feasible\ labeledroute$	$\#labeledroute$
$1 \rightarrow 2$	$(1,2,R)$	1
$1 \rightarrow 3$	$(1,2,R)(2,3,R)$	2
	$(1,2,R)(2,3,B)$	3
$1 \rightarrow 4$	$(1,2,R)(2,3,R)(3,4,R)$	4
	$(1,2,R)(2,3,B)(3,4,B)$	5
	$(1,2,R)(2,3,R)(3,4,G)$	6
	$(1,2,R)(2,3,B)(3,4,G)$	7
$1 \rightarrow 5$	$(1,2,R)(2,3,R)(3,4,R)(4,5,G)$	8
	$(1,2,R)(2,3,B)(3,4,B)(4,5,G)$	9
$2 \rightarrow 3$	$(2,3,R)$	10
	$(2,3,B)$	11
$2 \rightarrow 4$	$(2,3,R)(3,4,R)$	12
	$(2,3,B)(3,4,B)$	13
	$(2,3,R)(3,4,B)$	14
	$(2,3,R)(3,4,G)$	15
	$(2,3,B)(3,4,G)$	16
$2 \rightarrow 5$	$(2,3,R)(3,4,G)(4,5,G)$	17
	$(2,3,B)(3,4,G)(4,5,G)$	18
	$(2,3,R)(3,4,B)(4,5,G)$	19
$3 \rightarrow 4$	$(3,4,R)$	20
	$(3,4,B)$	21
	$(3,4,G)$	22
$3 \rightarrow 5$	$(3,4,G)(4,5,G)$	23
	$(3,4,R)(4,5,G)$	24
	$(3,4,B)(4,5,G)$	25
$4 \rightarrow 5$	$(4,5,G)$	26

Of course, we may have a more detailed representation, increasing the number of feasible labeled routes; e.g., considering the feasible labeled route labeled 14 in which the flow goes from node 2 to node 4, using first the arc $(2,4)$ labeled with R and then the arc $(3,4)$ labeled with B, it is possible to add another feasible labeled route using first the arc $(2,3)$ labeled with B and then the arc $(3,4)$ labeled with R. On the other hand, we may reduce the number of feasible labeled routes reducing the number of changes; e.g., referring to the OD $2 \rightarrow 4$, we may consider only the feasible labeled routes labeled 12 and 13, supposing that when flow goes on arcs with a particular label it does not change if it is not strictly needed.

Now, we can consider that the unit of flow is distributed among the feasible labeled routes as follows:

$\#labeled\ route$	1	2	3	4	5	6	7	8	9	10	11	12	13
$f(r)$	0.03	0.04	0.04	0.03	0.03	0.03	0.03	0.03	0.03	0.02	0.02	0.04	0.04

$\#labeled\ route$	14	15	16	17	18	19	20	21	22	23	24	25	26
$f(r)$	0.04	0.04	0.04	0.05	0.05	0.05	0.04	0.04	0.04	0.05	0.05	0.05	0.05

In view of function f, we can say that a 5% of the flow goes along labeled route 26, i.e., between nodes 4 and 5, or that labeled routes from 12 to 16 that connect nodes 2 and 4 are equally used.

Now, let us consider that there are several agents controlling different arcs of the network and labels are used to identify the agents who control the arcs of the network. In this sense, in Example 1 we would have up to three agents corresponding to the three labels R, B and G. For example, in a transport network these labels can represent different companies providing the passenger transport service between different cities or stops within the same city; or in a computer network these labels can represent links belonging to different Internet Service Providers (ISPs). If we are interested in knowing the relevance or contribution of each of the agents involved in the network, one possibility is to allocate the worth associated with the network in a *fair* way, i.e. determining which part of the worth associated with the network can be reasonably or in a fair way attributed to each agent. Formally, we introduce the following definition of labeled network allocation problem.

Definition 1. *Let $\mathcal{N} = (\mathcal{G}, \mathcal{R}, f)$, where $\mathcal{G} = (V, L, A)$, be a labeled network. A labeled network allocation problem associated with $\mathcal{N}$ is given by the 3-tuple $(\mathcal{N}, N, \mathcal{L})$ where:*

- $\mathcal{N}$ *is the labeled network.*
- $N = \{1, 2, \ldots, n\}$ *is the set of agents who control different arcs of the network.*
- $\mathcal{L} = \{L_1, L_2, ..., L_n\}$ *is a partition of the set of labels L, such that each agent $i \in N$ controls the subset L_i of labels.*

If each agent controls exactly one label, then the problem is called *simple labeled network allocation problem.*

We denote by $L(r)$ the set of different labels in labeled route r.

Example 2. *In order to illustrate the labeled network allocation problem, we refer to the situation in Example 1 where we consider that there are two agents, $\{A1, A2\}$ controlling labels $\{R, G\}$ and $\{B\}$ respectively. The next table shows the labeled routes controlled by each agent.*

Agent i	Feasible labeled route $r \in \mathcal{R} : L(r) \subseteq L_i$
A1	$1, 2, 4, 6, 8, 10, 12, 15, 17, 20, 22, 23, 24, 26$
A2	$11, 13, 21$

However, labeled routes $\{3, 5, 7, 9, 14, 16, 18, 19, 25\}$ need both agents to be completely controlled.

This labeled network allocation problem is not simple, because agent A1 controls two labels.

We are now interested in how to allocate the unit of flow among all agents controlling the different arcs of the network. This allocation give us the quotas or proportions of the unit of flow which are assigned or attributed to each agent.

These quotas will also measure, in some way, the contribution of each agent to the network.

Let $\mathcal{LNA}^N$ be the set of all labeled network allocation problems with set of agents N, a *flow quota allocation mechanism* for $\mathcal{LNA}^N$ is a function $\gamma : \mathcal{LNA}^N \to \mathbb{R}^N$ such that

1. $\gamma_i(\mathcal{N}, N, \mathcal{L}) \geq 0,\ \forall\, i \in N,$
2. $\sum_{i \in N} \gamma_i(\mathcal{N}, N, \mathcal{L}) = 1.$

Considering this definition, we can introduce many different quota allocation mechanisms. A simple possibility is the following. The flow of each feasible labeled route is divided equally among all agents involved. Summing up the results for agents, we obtain the amount of flow assigned to each agent. It is intuitive because the agents that are not involved in the flow of a given feasible labeled route do not take part in the division of the flow and those that are needed for determining the labeled route are rewarded equally, as each of them is equally important for that particular feasible labeled route. This procedure can be related to the well-known Shapley value (Shapley 1953) (see Algaba et al. 2019b), in the same way as in Algaba et al. (2019a).

Let $(\mathcal{N}, N, \mathcal{L})$ be a labeled network allocation problem, the *Shapley quota allocation mechanism* is defined for each $i \in N$ as follows:

$$\phi_i(\mathcal{N}, N, \mathcal{L}) = \sum_{r \in \mathcal{R}} \frac{f(r)}{\sum_{j \in N} \delta_j(L(r))} \delta_i(L(r)),$$

where $\delta_i(L(r)) = 1$, if $L(r) \cap L_i \neq \varnothing$, and $\delta_i(L(r)) = 0$, otherwise.

This flow quota allocation mechanism distributes the total flow 1 by labeled routes and within each labeled route equally among all agents involved in it. Furthermore, this allocation can be seen as a mixture of proportional allocation (of common revenue to the paths) and equal sharing (of path-revenue to providers). An alternative could be a proportional allocation of common revenue to the paths and proportional sharing of path-revenue to the arcs of each provider involved. Thus, we can define the *doubly proportional quota allocation mechanism* as follows:

$$\psi_i(\mathcal{N}, N, \mathcal{L}) = \sum_{r \in \mathcal{R}} \frac{f(r)}{\sum_{j \in N} \varepsilon_j(r)} \varepsilon_i(r),\ \ i \in N,$$

where $\varepsilon_i(r) = |\{(i_k, j_k, l_k) \in r : l_k \in L_i\}|$, i.e. the number of labeled arcs of route r whose labels belong to L_i.

Finally, we could also take into account different costs per route, different profits per route or different ticket prices by simply multiplying the function $f(r)$ in the numerator of the quota allocation mechanism by the corresponding cost, profit or ticket price.

3   Properties and Characterization

In this section, we deal with the problem of providing a set of properties that allow for the characterization of the Shapley quota allocation. In this context, we consider efficient solutions in order to allocate the full flow. The characterization of a solution is important because each proposed mechanism will give a different allocation of the flow. Consequently, it is not possible for all agents to agree on any solution. The axiomatic approach allows us to switch from a choice based on the amount each agent receives (or pays) to a choice focused on the fairness of the solution. Another positive aspect for studying the properties of a solution is that they can be used to explain the advantages of a solution more convincingly and so make it easier for the agents to accept. Some properties are the following:

- *No flow controlled property* (NFC): Let γ be a flow quota allocation mechanism defined on $\mathcal{LNA}^N$. It is said to satisfy the no flow controlled property, if for all $i \in N$, such that for all $r \in \mathcal{R}$, $L(r) \cap L_i = \varnothing$, then $\gamma_i = 0$.
- *Equal treatment of equals property* (ETE): Let γ be a flow quota allocation mechanism defined on $\mathcal{LNA}^N$. It is said to satisfy the equal treatment of equals property, if for all $i, j \in N$, such that for all $r \in \mathcal{R}$, $L(r) \cap L_i \neq \varnothing$ if and only if $L(r) \cap L_j \neq \varnothing$, then $\gamma_i = \gamma_j$.

The meaning of the no flow controlled property (NFC) is that agents which do not control any flow, will not be relevant to cooperation. The equal treatment of equals property (ETE) means that those agents which are symmetric, with respect to the number of labeled routes they participate in, must receive the same. Both properties seem reasonable and fair in the context of this problem.

An interesting question is how to merge two different labeled networks when both have the same set of labels and agents controlling the same labels. They could have different nodes, arcs and feasible labeled routes, but the merging of the two systems should provide a new labeled network involving all the structural elements of both. Additionally, one important aspect is that each system can have a different weight, relevance or size in terms of flow, so we should take this into account when merging both systems.

Let $(\mathcal{N}, N, \mathcal{L})$ and $(\mathcal{N}', N, \mathcal{L})$ be two labeled network allocation problems with set of agents N, and labeled networks $\mathcal{N} = (\mathcal{G}, \mathcal{R}, f)$ and $\mathcal{N}' = (\mathcal{G}', \mathcal{R}', f')$, such that $\mathcal{G} = (V, L, A)$ and $\mathcal{G}' = (V', L, A')$, and with relative weights w and w' ($w + w' = 1; w, w' > 0$), then a *merging* of both networks

$$(\mathcal{N}'', N, \mathcal{L}) \equiv (\mathcal{N}, N, \mathcal{L}) \oplus (\mathcal{N}', N, \mathcal{L})$$

is defined as follows:

- $\mathcal{G}'' = (V \cup V', L, A \cup A')$
- $\mathcal{R}' \cup \mathcal{R} \subseteq \mathcal{R}''$
- $f'' = wf + w'f'$ is given by

$$f''(r) = \begin{cases} wf(r) + w'f'(r), & \text{if } r \in \mathcal{R} \cap \mathcal{R}' \\ wf(r), & \text{if } r \in \mathcal{R} - \mathcal{R}' \\ w'f'(r), & \text{if } r \in \mathcal{R}' - \mathcal{R} \\ 0, & \text{if } r \in \mathcal{R}'' - (\mathcal{R} \cap \mathcal{R}') \end{cases} \qquad \forall\, r \in \mathcal{R}''$$

It is not difficult to check that all elements are well-defined[1]. Likewise, $f''(r) = 0, \quad \forall r \in \mathcal{R}''(A'')\backslash\mathcal{R}''$. Therefore, the definition of f'' implies that new possible labeled routes do not generate new flow when merging the labeled networks.

The interpretation of the merging operation is that we construct a new labeled network whose graph structure consists of all the structural elements of both graphs and the weights are used to redefine the distribution of the unit of flow adapted to the new structure.

We now introduce the following property for solutions in labeled networks with set of agents N.

- *Weighted merging property* (WM): Let $(\mathcal{N}, N, \mathcal{L})$ and $(\mathcal{N}', N, \mathcal{L})$ be two labeled network allocation problems with set of agents N, and labeled networks $\mathcal{N} = (\mathcal{G}, \mathcal{R}, f)$ and $\mathcal{N}' = (\mathcal{G}', \mathcal{R}', f')$, such that $\mathcal{G} = (V, L, A)$ and $\mathcal{G}' = (V', L, A')$, and with relative weights w and w'. And let γ be a flow quota allocation mechanism defined on $\mathcal{LNA}^N$. It is said to satisfy the weighted merging property, if the following holds

$$\gamma((\mathcal{N}, N, \mathcal{L}) \oplus (\mathcal{N}', N, \mathcal{L})) = w\gamma(\mathcal{N}, N, \mathcal{L}) + w'\gamma(\mathcal{N}', N, \mathcal{L}).$$

Proposition 1. *The Shapley quota allocation mechanism satisfies NFC, ETE and WM.*

Proof. It is straightforward to prove that the Shapley quota allocation mechanism satisfies NFC and ETE. Let $(\mathcal{N}, N, \mathcal{L})$ and $(\mathcal{N}', N, \mathcal{L})$ be two labeled network allocation problems with set of agents N, and labeled networks $\mathcal{N} = (\mathcal{G}, \mathcal{R}, f)$ and $\mathcal{N}' = (\mathcal{G}', \mathcal{R}', f')$, such that $\mathcal{G} = (V, L, A)$ and $\mathcal{G}' = (V', L, A')$, and with relative weights w and w'. On the one hand, we have for every $i \in N$

$$\phi_i(\mathcal{N}, N, \mathcal{L}) = \sum_{r \in \mathcal{R}} \frac{f(r)}{\sum_{j \in N} \delta_j(L(r))} \delta_i(L(r)),$$

and

$$\phi_i(\mathcal{N}', N, \mathcal{L}) = \sum_{r \in \mathcal{R}'} \frac{f'(r)}{\sum_{j \in N} \delta_j(L(r))} \delta_i(L(r)).$$

On the other hand, for every $(\mathcal{N}'', N, \mathcal{L}) \equiv (\mathcal{N}, N, \mathcal{L}) \oplus (\mathcal{N}', N, \mathcal{L})$, we have that

$$\phi_i(\mathcal{N}'', N, \mathcal{L}) = \sum_{r \in \mathcal{R}''} \frac{f''(r)}{\sum_{j \in N} \delta_j(L(r))} \delta_i(L(r))$$

---

[1] The difference of two sets A and B in the definition of $f''(r)$ is as follows: $A - B = A \backslash (A \cap B)$.

$$= \sum_{r \in \mathcal{R} \cap \mathcal{R}'} \frac{wf(r) + w'f'(r)}{\sum_{j \in N} \delta_j(L(r))} \delta_i(L(r)) + \sum_{r \in \mathcal{R} - \mathcal{R}'} \frac{wf(r)}{\sum_{j \in N} \delta_j(L(r))} \delta_i(L(r))$$

$$+ \sum_{r \in \mathcal{R}' - \mathcal{R}} \frac{w'f'(r)}{\sum_{j \in N} \delta_j(L(r))} \delta_i(L(r))$$

$$= \sum_{r \in \mathcal{R}} \frac{wf(r)}{\sum_{j \in N} \delta_j(L(r))} \delta_i(L(r)) + \sum_{r \in \mathcal{R}'} \frac{w'f'(r)}{\sum_{j \in N} \delta_j(L(r))} \delta_i(L(r))$$

$$= w \sum_{r \in \mathcal{R}} \frac{f(r)}{\sum_{j \in N} \delta_j(L(r))} \delta_i(L(r)) + w' \sum_{r \in \mathcal{R}'} \frac{f'(r)}{\sum_{j \in N} \delta_j(L(r))} \delta_i(L(r))$$

$$= w \phi_i(\mathcal{N}, N, \mathcal{L}) + w' \phi_i(\mathcal{N}', N, \mathcal{L}).$$

Therefore, the statement holds.   □

Theorem 1. *The Shapley quota allocation mechanism is the unique flow quota allocation mechanism satisfying NFC, ETE and WM on $\mathcal{LNA}^N$.*

Proof. Let $(\mathcal{N}, N, \mathcal{L})$ be a labeled network allocation problem with set of agents N, labeled networks $\mathcal{N} = (\mathcal{G}, \mathcal{R}, f)$, such that $\mathcal{G} = (V, L, A)$ and let γ be a quota solution satisfying NFC, ETE and WM on $\mathcal{LNA}^N$.

Let $|\mathcal{R}| = 1$, i.e., $\mathcal{R}$ contains a single labeled route r and $f(r) = 1$. Since γ satisfies NFC we have that for all i such that $L(r) \cap L_i = \varnothing$, $\gamma_i(\mathcal{N}, N, \mathcal{L}) = 0$. Now as γ satisfies ETE, we obtain that for all i such that $L(r) \cap L_i \neq \varnothing$, $\gamma_i(\mathcal{N}, N, \mathcal{L}) = \frac{1}{\sum_{j \in N} \delta_j(L(r))}$. Now by definition of the Shapley quota allocation mechanism ϕ, we have that

$$\phi_i(\mathcal{N}, N, \mathcal{L}) = \begin{cases} \frac{1}{\sum_{j \in N} \delta_j(L(r))}, & \text{if } L(r) \cap L_i \neq \emptyset \\ 0, & \text{if } L(r) \cap L_i = \emptyset \end{cases}, \quad \forall i \in N.$$

Therefore, $\gamma = \phi$.

Let us suppose by induction that $\gamma = \phi$ for every $(\mathcal{N}, N, \mathcal{L}) \in \mathcal{LNA}^N$ such that $|\mathcal{R}| \leq m - 1, m > 1$, and let us consider $(N, \mathcal{G}, \mathcal{R}, f) \in \mathcal{LN}^N$ such that $|\mathcal{R}| = m > 1$.

We choose one labeled route $r \in \mathcal{R}$, and we construct the two following labeled networks with set of agents N:

- $\mathcal{N}^1 = (\mathcal{G}^1, \mathcal{R}^1, f^1)$:
  1. $\mathcal{G}^1 = \mathcal{G}$.
  2. $\mathcal{R}^1 = \mathcal{R} - \{r\}$.
  3. $f^1(s) = \dfrac{f(s)}{1 - f(r)}, \ \forall \, s \in \mathcal{R}^1$.

  Since $|\mathcal{R}| \geq 2$, $p^1(s)$ is well-defined because $0 < f(r) < 1$.
- $\mathcal{N}^2 = (\mathcal{G}^2, \mathcal{R}^2, f^2)$:
  1. $\mathcal{G}^2 = \mathcal{G}$.
  2. $\mathcal{R}^2 = \{r\}$.
  3. $f^2(r) = 1, \ r \in \mathcal{R}^2$.

If we consider the merging of $(\mathcal{N}^1, N, \mathcal{L})$ and $(\mathcal{N}^2, N, \mathcal{L})$ with $\mathcal{R} = \mathcal{R}^1 \cup \mathcal{R}^2$ and relative weights $1 - f(r)$ and $f(r)$ respectively, then applying the definition of merging of two labeled networks we obtain that

$$(\mathcal{N}, N, \mathcal{L}) \equiv (\mathcal{N}^1, N, \mathcal{L}) \oplus (\mathcal{N}^2, N, \mathcal{L}).$$

Now, by the induction hypothesis, we have that

$$\gamma(\mathcal{N}^1, N, \mathcal{L}) = \phi(\mathcal{N}^1, N, \mathcal{L}),$$

$$\gamma(\mathcal{N}^2, N, \mathcal{L}) = \phi(\mathcal{N}^2, N, \mathcal{L}).$$

By Proposition 1, ϕ satisfies WM and by hypothesis γ satisfies WM, then we have the following chain of equalities:

$$\gamma(\mathcal{N}, N, \mathcal{L}) = \gamma((\mathcal{N}^1, N, \mathcal{L}) \oplus (\mathcal{N}^2, N, \mathcal{L})) = (1 - f(r))\gamma(\mathcal{N}^1, N, \mathcal{L}) + f(r)\gamma(\mathcal{N}^2, N, \mathcal{L})$$
$$= (1 - f(r))\phi(\mathcal{N}^1, N, \mathcal{L}) + f(r)\phi(\mathcal{N}^2, N, \mathcal{L}) = \phi((\mathcal{N}^1, N, \mathcal{L}) \oplus (\mathcal{N}^2, N, \mathcal{L}))$$
$$= \phi(\mathcal{N}, N, \mathcal{L}).$$

Therefore $\gamma = \phi$, and the result follows.   □

Theorem 2. *The properties NFC, ETE and WM are logically independent.*

Proof. (1) The egalitarian solution satisfies ETE and WM but not NFC. Indeed, the egalitarian solution is defined as follows:

$$\varepsilon_i(\mathcal{N}, N, \mathcal{L}) = \frac{1}{|N|}, \quad \forall\, i \in N.$$

It can immediately be proved that the egalitarian solution does not satisfy NFC. Furthermore, it trivially satisfies ETE because all agents receive the same. Since the egalitarian solution is a constant solution on $\mathcal{LNA}^N$, it also satisfies WM.

(2) Let us consider the following version of the egalitarian solution:

$$\alpha_i(\mathcal{N}, N, \mathcal{L}) = \begin{cases} \frac{1}{|K|}, & \text{if } i \in K \subseteq N \\ 0, & \text{otherwise} \end{cases}, \quad \forall\, i \in N,$$

where $K = \{i \in N : \exists\, r \in \mathcal{R} \text{ s.t. } L(r) \cap L_i \neq \emptyset\}$. It is easy to prove that this solution satisfies NFC and ETE. However, it does not satisfy WM. Let us consider the following two labeled networks with set of agents $N = \{1, 2, 3, 4, 5\}$ and $\mathcal{L} = \{\{1\}, \{2\}, \{3\}, \{4\}, \{5\}\}$:

- $\mathcal{N}^1 = (\mathcal{G}, \mathcal{R}^1, f^1) : \mathcal{R}^1 = \{r\}; L(r) = \{1, 2, 3\}; f^1(r) = 1,$
- $\mathcal{N}^2 = (\mathcal{G}, \mathcal{R}^2, f^2) : \mathcal{R}^2 = \{s\}; L(s) = \{4, 5\}; f^2(s) = 1,$

with relative weights equal to $\frac{1}{2}$. Then we have that

$$\alpha((\mathcal{N}^1, N, \mathcal{L}) \oplus (\mathcal{N}^2, N, \mathcal{L})) = \left(\frac{1}{5}, \frac{1}{5}, \frac{1}{5}, \frac{1}{5}, \frac{1}{5}\right),$$

while $\alpha(\mathcal{N}^1, N, \mathcal{L}) = \left(\frac{1}{3}, \frac{1}{3}, \frac{1}{3}, 0, 0\right)$ and $\alpha(\mathcal{N}^2, N, \mathcal{L}) = \left(0, 0, 0, \frac{1}{2}, \frac{1}{2}\right).$

Thus, $\frac{1}{2}\left(\frac{1}{3}, \frac{1}{3}, \frac{1}{3}, 0, 0\right) + \frac{1}{2}\left(0, 0, 0, \frac{1}{2}, \frac{1}{2}\right) = \left(\frac{1}{6}, \frac{1}{6}, \frac{1}{6}, \frac{1}{4}, \frac{1}{4}\right).$ Hence α does not satisfies WM.

(3) Let us consider the solution defined, for each $i \in N$, as follows:

$$\varphi_i(\mathcal{N}, N, \mathcal{L}) = \sum_{r \in \mathcal{R}} \frac{i \cdot f(r)}{\sum_{j \in N} j \cdot \delta_j(L(r))} \delta_i(L(r)).$$

This solution satisfies NFC and also WM but not ETE.   □

4   A Stylized Application to the Metropolitan Consortium of Seville

The purpose of this section is to illustrate how labeled networks can describe real-life situations such as a public transport system. Furthermore, the Shapley quota allocation mechanism is not difficult to compute and is easy to apply, because we only need to know the labeled routes and the distribution of one unit of flow among all labeled routes. Of course, in real-life situations both elements should be updated from time to time.

In particular, we apply labeled networks and the Shapley quota allocation mechanism to the Transport Consortium of Seville in which collaborate several transport companies to offer a better service to the passengers, in particular the Consortium offers travel tickets which can be used in whatever of the transport companies including transfers between the different companies. This transport system covers six zones (A,...,F) and connects different points of Seville and its metropolitan area. This network uses three modes: trains, metro and buses. In fact, there are 5 lines operated by trains, 3 by underground and 64 by buses which correspond to seven different companies. The complete map can be found at the web page www.consorciotransportes-sevilla.com (see also Fig. 2). We should mention that the urban buses are not included in the transport consortium of Seville.

A public transport system as described above can be modeled by means of a simple labeled network allocation problem as follows:

- The labeled graph $\mathcal{G}$:
    - $V = \{$ the set of all stops $\}$.
    - $L = \{$ the set of all companies operating in the transport system $\}$.
    - $A = \{$ each connection between two consecutive stops operated by each transport company $\}$.

- The set of agents $N = \{$ the set of all companies operating in the transport system $\}$.
- The partition of the labels, $\mathcal{L} = \{\{\text{company } 1\}, \{\text{company } 2\}, \ldots, \{\text{company } n\}\}$.
- The feasible labeled routes $\mathcal{R} = \{$ the set of all labeled routes used by passengers $\}$.
- The distribution of one unit of the flow $f(r) = $ proportion of all passengers using *labeled route r*.

If we consider that the price of one ticket is constant, which is true within the same zone, then we can consider that this price is exactly 1, and f measures the proportion of profit derived of the use of each labeled route. Thus, the application of a flow quota allocation mechanism provides the proportion of the ticket price that corresponds to each agent when they cooperate.

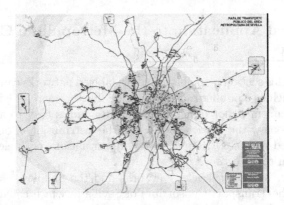

Fig. 2. Map of the metropolitan consortium of Seville

The real network is too large to be illustrated in the paper, so we have decided to give our attention to a limited problem in which we consider only zone A which corresponds to the city center of Seville. Moreover, we aggregate different stops, if they serve the same area and different lines, if they have common labeled routes and common stops in the area under consideration. We do not consider those companies that do not operate at all in the area or if the service they provide is limited to few stops. This leads us to consider three bus companies, Carjema (B1), Casal (B2), and Los Amarillos (B3); underground, Metro (M), and trains, Renfe (T). The resulting network has 15 nodes and 15 edges some of which are connected by different companies (see Fig. 3). All these simplifications do not affect the way in which the Shapley quota allocation mechanism would be applied, therefore, this example adequately illustrates how this mechanism would be calculated in a real-life situation. In fact, the computational complexity of the problem lies in the algorithm for determining all possible routes and not

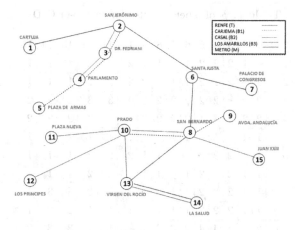

Fig. 3. Simplified map of the Zone A of the metropolitan network of Seville

in the application of the quota allocation mechanism. But the first is not the subject of this paper.

Starting from the simplified network, we compute the feasible labeled routes according to the following hypotheses: a passenger enters the origin node and takes the first public transport available traveling towards the destination. The passenger remains on this public transport as long as possible. When the passenger leaves that public transport s/he then takes again the first public transport available going to the destination, and so on until reaching the final destination. Following this procedure, we obtained a total of 405 feasible labeled routes (see Table 1). The number of labeled routes operated by each company are shown in Table 2

For instance, let us consider a passenger who wants to go from Plaza de Armas (node 5) to San Jerónimo (node 2). We assume that there is just one feasible labeled route, because in Plaza de Armas the only transport is by Casal company (B2) bus that goes directly to the final destination of San Jerónimo. In other words, we do not consider as feasible the labeled route that uses the B2 bus from Plaza de Armas to Parlamento (node 4) and the Carjema company (B1) bus from Parlamento to San Jerónimo, and the labeled route that uses the B2 bus from Plaza de Armas to Dr. Fedriani (node 3) and B1 bus from Dr. Fedriani to San Jerónimo. On the other hand, for the path in the opposite direction, from San Jerónimo to Plaza de Armas, the direct labeled route with B2 bus and the labeled route that uses B1 bus from San Jerónimo to Parlamento and B2 bus from Parlamento to Plaza de Armas, both are feasible, depending on which bus arrives first at San Jerónimo, but we do not consider as feasible the labeled route that uses the B1 bus from San Jerónimo to Dr. Fedriani and the B2 bus from Dr. Fedriani to Plaza de Armas because Dr. Fedriani is not the last possible stop for bus B1.

Table 1. # of labeled routes for each pair O-D

OD	1	2	3	4	5	6	7	8	9	10	11	12	13	14	15
1	–	1	2	2	2	1	1	1	1	2	2	2	1	1	1
2	1	–	2	2	2	1	1	1	1	2	2	2	1	1	1
3	2	2	–	2	2	2	2	2	2	4	3	3	2	2	2
4	2	2	2	–	1	2	2	2	2	4	3	3	2	2	2
5	1	1	1	1	–	1	1	1	1	2	1	1	1	1	1
6	1	1	2	2	2	–	1	1	1	2	2	2	1	1	1
7	1	1	2	2	2	1	–	1	1	2	2	2	1	1	1
8	1	1	2	2	2	1	1	–	1	2	2	2	3	3	1
9	1	1	2	2	2	1	1	1	–	1	1	1	2	2	1
10	2	2	4	4	4	2	2	2	2	–	1	1	3	3	2
11	1	1	2	2	2	1	1	1	1	1	–	1	2	2	1
12	1	1	2	2	2	1	1	1	1	1	1	–	2	2	1
13	3	3	6	6	6	3	3	3	3	3	3	3	–	2	3
14	4	4	7	7	8	4	4	4	4	4	3	4	2	–	4
15	1	1	2	2	2	1	1	1	1	1	1	1	2	2	–

Table 2. The labeled routes and the companies

Companies	M	T	B1	B2	B3
# of labeled routes involving each company	180	315	89	203	88
# of labeled routes operated by a unique company	20	42	6	18	6

After the previous modifications to the network, we have simulated the O-D matrix, taking into consideration the population distribution in the metropolitan area and available data of passengers in 2014.

Based on data provided by the Metropolitan Consortium of Seville, we have assigned a weight to each node that represents its relevance in the traffic (see Table 3) and consider an average of 13.500.000 passengers per year. We have simulated the O-D matrix, where the average of passengers in each O-D has been calculated proportional to the products of the assigned weights to each node in the pair O-D. For each pair O-D, we apply a normal distribution with the average previously obtained and a relative standard deviation (coefficient of variation) of 7.5%. Next, we have generated random numbers for the matrix O-D. Using these numbers and the normal distribution we have obtained a traffic flow throughout the transport network (see Table 4). We would like to emphasize that we have implemented a program on spreadsheets to simulate the traffic of passengers taking as input the stop weights, the relative standard deviation and the average of total passengers in order to calculate the Shapley quota allocation mechanism and a proportional solution according to the number of

Table 3. Weights of the nodes

Node	1	2	3	4	5	6	7	8	9	10	11	12	13	14	15
Weight	1	3	3	3	10	8	2	8	2	10	5	5	8	4	2

Table 4. Number of passengers for each pair O-D

OD	1	2	3	4	5	6	7	8	9	10	11	12	13	14	15
1	–	7904	8771	8176	29245	22734	5617	20740	5306	25433	13508	14606	20544	9116	5135
2	8145	–	23641	25489	85749	64302	14529	61720	13393	75368	36430	39825	66047	32446	16654
3	7102	19765	–	22741	84691	69579	16014	62701	17586	82785	38789	48299	65673	30181	15968
4	8969	24321	23690	–	81794	56529	17777	71428	16933	84858	43108	41902	65342	28700	17068
5	25786	89615	89606	76253	–	222254	57499	218904	49359	274017	138958	145169	218926	103733	46576
6	20761	64905	64244	66932	221379	–	41907	157638	47267	214889	115142	112964	158628	80630	43869
7	5110	16674	18278	16230	54424	43022	–	40096	10441	60367	30912	31284	36590	21881	10431
8	22536	55140	64589	62622	237452	173367	42101	–	44490	249182	108662	120792	175477	81645	40405
9	6408	16724	17832	16488	52697	42572	12117	50059	–	61691	30058	27738	44548	21937	10662
10	26936	67826	78332	91698	305185	207276	59092	207887	58424	–	150398	138593	214263	106047	55600
11	12315	41216	44813	42168	144143	107359	28061	116516	30423	118116	–	63994	105589	54928	30030
12	15915	37891	38309	40060	138808	103039	25341	106975	29109	128213	65672	–	103663	48972	28294
13	23586	61210	57843	59137	214729	190465	45843	193877	34290	202610	108369	109267	–	89055	45269
14	11006	31567	33569	30413	95339	86084	24804	94003	21720	127600	60320	54098	89252	–	24125
15	5139	16814	15119	16717	51685	48327	10831	42964	9987	58279	26595	27805	37667	21015	–

labeled routes. We would like to point out that if a real traffic matrix is available then it can be easily imported.

In our problem the number of labeled routes is fixed. Notice that this is not a strong assumption since the licenses are conceded to the companies for a long period of time. Although the computation of the number of feasible labeled routes is an NP-hard problem, this is initially solved by the metropolitan consortium and it is beyond the scope of this paper.

Once simulation is applied to the previous data, the Shapley quota allocation mechanism is obtained together with a proportional solution based on the number of labeled routes in which every company operates (see Table 2). These solutions, as a percentage of the total price of the ticket, are reported in Tables 5 and 6.

Table 5. The Shapley quota allocation mechanism

M	T	B1	B2	B3
24.98	37.82	5.34	25.17	6.69

In Tables 5 and 6, we can observe that the results are quite different. For instance, Company B1 would obtain a quota of 5.34% with the Shapley quota allocation mechanism and 10.17% considering the proportional solution. We

Table 6. Proportional quota according to the number of labeled routes

M	T	B1	B2	B3
20.57	36.00	10.17	23.20	10.06

highlight that, whereas the proportional quota only takes into account the number of labeled routes, the Shapley quota allocation mechanism is based not only on the number of labeled routes, but also on the traffic flow in the network. In this case, this solution is also easily computed which is a relevant advantage when working with examples from the real world and has a clear interpretation. Note that we have assumed in this case, that if there is more than one labeled route connecting the same pair OD, we consider them equally probable. However, different probabilities could have been considered as shown in the theoretical part. In that case, only one additional input should be added to the program which can easily be implemented.

5   Conclusions

In this paper, we have introduced a class of network problems, the labeled network allocation problems. Furthermore, we have studied and characterized the so-called Shapley quota allocation mechanism, which is based on the structure of the Shapley value but directly used with the labeled network.

Finally, we have illustrated the application of labeled network models to real-life situations by using a stylized example of the Metropolitan Consortium of Seville.

Acknowledgements. First of all, the authors thank two anonymous referees for her/his helpful comments and suggestions to improve the contents of the paper. Financial support from the Ministerio de Ciencia, Innovación y Universidades (MCIU), the Agencia Estatal de Investigación (AEI) and the Fondo Europeo de Desarrollo Regional (FEDER) under the project PGC2018-097965-B-I00 is gratefully acknowledged.

References

Algaba, E., Fragnelli, V., Llorca, N., Sánchez-Soriano, J.: Horizontal cooperation in a multimodal public transport system: the profit allocation problem. Eur. J. Oper. Res. **275**, 659–665 (2019a)

Algaba, E., Fragnelli, V., Sánchez-Soriano, J.: Handbook of the Shapley Value, CRC Press, Taylor & Francis Group, Boca Raton (2019b). ISBN 9780815374688

Algaba, E., van den Brink, R., Dietz, C.: Power measures and solutions for games under precedence constraints. J. Optim. Theory Appl. **172**, 1008–1022 (2017)

Algaba, E., van den Brink, R., Dietz, C.: Network structures with hierarchy and communication. J. Optim. Theory Appl. **179**, 265–282 (2018)

Ball, M.O., Magnanti, T.L., Monma, C.L., Nemhauser, G.L.: Network models. In: Handbooks in Operations Research and Management Science, vol. 7. North-Holland, Amsterdam (1995a)

Ball, M.O., Magnanti, T.L., Monma, C.L., Nemhauser, G.L.: Network routing. In: Handbooks in Operations Research and Management Science, vol. 8. North-Holland, Amsterdam (1995b)

Bergantiños, G., Gómez-Rúa, M., Llorca, N., Pulido, M., Sánchez-Soriano, J.: A new rule for source connection problems. Eur. J. Oper. Res. **234**, 780–788 (2014)

Bjørndal, E., Koster, M., Tijs, S.: Weighted allocation rules for standard fixed tree games. Math. Methods Oper. Res. **59**, 249–270 (2004)

Curiel, I., Derks, J., Tijs, S.: On balanced games and games with committee control. OR Spektrum **11**, 83–88 (1989)

Derks, J.J.M., Tijs, S.H.: Stable outcomes for multi-commodity flow games. Methods Oper. Res. **50**, 493–504 (1985)

Derks, J.J.M., Tijs, S.H.: Totally balanced multi-commodity games and flow games. Methods Oper. Res. **54**, 335–347 (1986)

Fragnelli, V., García-Jurado, I., Norde, H., Patrone, F., Tijs, S.: How to share railway infrastructure costs? In: Patrone, F., Tijs, S. (eds.) Game Practice: Contributions from Applied Game Theory (García-Jurado I, pp. 91–101. Kluwer, Amsterdam (2000a)

Fragnelli, V., García-Jurado, I., Méndez-Naya, L.: On shortest path games. Math. Methods Oper. Res. **52**, 251–264 (2000b)

Granot, D., Hojati, M.: On cost allocation in communication networks. Networks **20**, 209–229 (1990)

Gupta, A., Srinivasan, A., Tardos, É.: Cost-sharing mechanisms for network design. In: Jansen, K., Khanna, S., Rolim, J.D.P., Ron, D. (eds.) APPROX/RANDOM -2004. LNCS, vol. 3122, pp. 139–150. Springer, Heidelberg (2004). https://doi.org/10.1007/978-3-540-27821-4_13

Jackson, M., Zenou, Y.: Games on networks. In: Zamir, S., Young, P. (eds.) Handbook of Game Theory, vol. 4, Chapter 3, 89 pages. North Holland (Elsevier), Amsterdam (2014)

Kalai, E., Zemel, E.: Totally balanced games and games of flow. Math. Oper. Res. **7**, 476–478 (1982)

Koster, M., Molina, E., Sprumont, Y., Tijs, S.: Sharing the cost of a network: core and core allocations. Int. J. Game Theory **30**, 567–599 (2001)

Maschler, M., Potters, J., Reijnierse, H.: The nucleolus of a standard tree game revisited: a study of its monotonicity and computational properties. Int. J. Game Theory **39**, 89–104 (2010)

Megiddo, N.: Cost allocation for steiner trees. Networks **8**, 1–6 (1978)

Norde, H., Fragnelli, V., García-Jurado, I., Patrone, F., Tijs, S.: Balancedness of infrastructure cost games. Eur. J. Oper. Res. **136**, 635–654 (2002)

Reijnierse, J.H., Maschler, M.B., Potters, J.A.M., Tijs, S.H.: Simple flow games. Games Econ. Behav. **16**, 238–260 (1996)

Roughgarden, T., Schrijvers, O.: Network cost-sharing without anonymity. In: Lavi, R. (ed.) SAGT 2014. LNCS, vol. 8768, pp. 134–145. Springer, Heidelberg (2014). https://doi.org/10.1007/978-3-662-44803-8_12

Sánchez-Soriano, J.: The pairwise egalitarian solution. Eur. J. Oper. Res. **150**, 220–231 (2003)

Sánchez-Soriano, J.: Pairwise solutions and the core of transportation situations. Eur. J. Oper. Res. **175**, 101–110 (2006)

Sánchez-Soriano, J., Llorca, N., Meca, A., Molina, E., Pulido, M.: An integrated transport system for Alacant's students. UNIVERCITY, Ann. Oper. Res. **109**, 41–60 (2002)

Shapley, L.S.: A value for n-person games. Ann. Math. Stud. **28**, 307–317 (1953)

Sharkey, W.W.: Network models in economics. In: Ball, M.O., Magnanti, T.L., Monma, C.L., Nemhauser, G.L. (eds.) Handbooks in Operations Research and Management Science, vol. 8, pp. 713–765 (1995)

Slikker, M., van den Nouweland, A.: Social and Economic Networks in Cooperative Game Theory. Kluwer Academic Publisher, Norwell (2001)

Tardos, E.: Network games. In: STOC 2004 Proceedings of the Thirty-Sixth Annual ACM Symposium on Theory of Computing, pp. 341–342 (2004)

Tijs, S., Moretti, S., Branzei, R., Norde, H.: The bird core for minimum cost spanning tree problems revisited: monotonicity and additivity aspects. In: Seeger, A. (ed.) Recent Advances in Optimization. Lecture Notes in Economics and Mathematical Systems, vol. 563, pp. 305–322. Springer, Heidelberg (2006). https://doi.org/10.1007/3-540-28258-0_19

Seat Apportionment by Population and Contribution in European Parliament After Brexit

Cesarino Bertini[1] , Gianfranco Gambarelli[1] ,
Izabella Stach[2(✉)] , and Giuliana Zibetti[3]

[1] Department of Management, Economics and Quantitative Methods, University
of Bergamo, Via dei Caniana, 2, 24127 Bergamo, Italy
{cesarino.bertini,gianfranco.gambarelli}@unibg.it
[2] AGH University of Science and Technology, Al. Mickiewicza 30,
30-059 Krakow, Poland
istach@zarz.agh.edu.pl
[3] Institute Leonardo da Vinci, Bergamo, Italy
g.zibetti@datamorfosi.com

Abstract. The problem of apportioning seats to member countries of the European Parliament after Brexit and in view of new accessions/exits is delicate, as countries with strong economies (and their consequent large contributions to the European Union) require that they have greater representative weight in the European Parliament. In this paper, we propose a model for seat apportionment in the European Parliament, which assigns seats taking into account both the percentages of the populations and the percentages of the contributions by each member state to the European Union budget by means of a linear combination of these two quantities. The proposed model is a modification of the approach given by Bertini, Gambarelli, and Stach in 2005. Using the new model, we studied the power position of each European Union member state before and after the exit of the United Kingdom using the Banzhaf power index. A short latest-literature review on this topic is given.

Keywords: Apportionments · Brexit · Cooperative game theory · European Parliament · Power indices

1 Introduction

The European Parliament (EP) is made up of 751 members elected by the 28 member states of the European Union (EU). The United Kingdom (UK) started the process of leaving the European Union (the so-called Brexit) on March 29, 2017, and should have left it on March 29, 2019, which did not happen. According to current agreement between the UK and the EU, the new deadline for Brexit is fixed on October 31, 2019. The Brexit extension allows for the ratification of the withdrawal agreement. After Brexit, the situation will change. The EP proposition is that the number of members of the European Parliament (MEPs) will change from 751 to 705 after Brexit (see Table 1). The general assumption is that none of the remaining EU countries will lose

© Springer-Verlag GmbH Germany, part of Springer Nature 2019
N. T. Nguyen et al. (Eds.): TCCI XXXIV, LNCS 11890, pp. 109–126, 2019.
https://doi.org/10.1007/978-3-662-60555-4_8

seats; however, the problem of apportioning seats to member countries of the European Parliament after Brexit and in view of new accessions/exits is delicate, as countries with strong economies (and their consequent large contributions to the European Union) require that they have greater representative weight in the EP. Naturally, this is a very questionable and debated political issue that clashes with the idea of solidarity at the base of every community.

Without entering into "politically correct" or "politically incorrect" aspects in this paper, we propose a method for seat apportionment in the EP. This method is a modification of the approach introduced by Bertini, Gambarelli, and Stach in [1] and then applied in [2]. The new method assigns seats to the member states of EU by taking into account both the populations and contributions of each member state to the European Union budget by means of a linear combination of these two quantities. In an old model proposed in 2005, gross domestic product (GDP) was considered instead of EU contributions. Using the new model, we study the power position of each EU member state before and after the exit of the UK. In our research, we used the normalized Banzhaf power index in order to measure the influence of each member state in the EP. In order to calculate the Banzhaf power index efficiently, we applied the combinatorial method based in generating functions introduced by Bilbao, Fernández, Jiménez Losada, and López in [3].

The effect of Brexit on the power relationships and measuring power in the European Union decision-making bodies is of great interest to different researchers. In particular, how the distribution of power in the EP and the Council of the EU might change (if at all) is an interesting question. Let us briefly present some of the latest publications on this topic, which present a cooperative-game-theoretical approach to this issue. Mercik and Ramsey analyzed the changes in the power relationships in the Council of EU (Council) after Brexit in [4]. Their approach to measure the voting power of each EU country is based on the concept of pre-coalitions that are formed around the most populated countries. They considered adaptations of three power indices: the Shapley-Shubik, Banzhaf and Johnston indices based on the so-called quotient index (for Shapley-Shubik, see [5] and [6]; for Banzhaf, see [7] and Sect. 2; for Johnston, see [8], for example). One of the surprising conclusions of their research is that the exit of the UK will effectively decrease the power of the small countries. Then, they (along with Gładysz) continued the research on the effect of Brexit on the balance of power in the Council in [9] by applying a fuzzy multicriteria attempt. For this purpose, they proposed a fuzzy definition of the classic Shapley-Shubik power index. They also provided a comparison of the pre- and post-Brexit distribution of power in the standard and fuzzy approaches. In [10], Kóczy started to analyze the effect of the withdrawal of the UK from the EU on the distribution of power in the in the Council. Then, along with Petróczy and Rogers, he continued an examination into what would happen in the Council after not only Brexit but the potential exit of any other individual country from the EU in [11]. For this purpose, they measured the potential changes in the influence of each country using the Shapley-Shubik index. In addition, this approach confirmed that Brexit (as well as any other potential withdrawal from the 28-member European Union) increases the power of the countries with the largest populations. Moreover, they analyzed a hypothetical Brexit before the accession of

Croatia, obtaining that it would diminish the power of the greater-populated countries in the Council.

The structure of this paper is as follows. In the following Sect. 2, we introduce the preliminary definitions and notations of the cooperative game theory necessary to better understand the method proposed. Section 3 provides a short description of the European Parliament. In Sect. 4, the seat apportionment method proposed by Gambarelli, Bertini, and Stach in 2005 is mentioned, and our new method is given. Section 5 is devoted to the application of new method to the European Parliament before and after the exit of the United Kingdom. Section 6 concludes the paper with a proposition of further developments as well as our concluding remarks.

2 Preliminary Definitions and Notations

The voting situation in the EP can be described as a weighted majority game (i.e., a specific kind of cooperative game). Hereafter, we will give all of the necessary definitions and notations that refer to cooperative games and the measurement of the power of players in voting situations.

A *cooperative game in its characteristic form* is defined as pair (N, v) where N is a finite set of players, 2^N is a set of all possible subsets (i.e., coalitions of players), and $v : 2^N \rightarrow R$ is a function that assigns $S \subseteq N$ a worth of $v(S)$ to every coalition with a requirement that empty coalition $\emptyset$ has no worth (i.e., $v(\emptyset) = 0$). Let (N, v) be a cooperative game; game (N, v) is said to be monotonic if $v(S) \leq v(T)$ whenever $S \subseteq T$ for all $S, T \subseteq N$. Game (N, v) is said to be a *simple game* if v has values in set $\{0, 1\}$ for every coalition S in N. If $v(S) = 1$, then S is called a *winning coalition*; otherwise, S is called a *losing coalition*. Simple game (N, v) is said to be a *weighted majority game* if there exist real weights $w_i \geq 0$ for every $i \in N$ and non-negative quota q, $\frac{1}{2} \sum_{i \in N} w_i < q \leq \sum_{i \in N} w_i$ such that $v(S) = 1$ if and only if $\sum_{i \in S} w_i \geq q$. If in a vote $q = \left\lfloor \frac{1}{2} \sum_{i \in N} w_i \right\rfloor + 1$ we talk about so-called *simple majority rule* ($\lfloor x \rfloor$ denotes the integral part of x, i.e., the greatest integer not greater than x).

We denote the cardinality of coalition S by $|S|$ and, specifically, we put $n = |N|$.

Power index f on the sets of all simple games is a map that assigns unique vector $f(v) = (f_1(v), \ldots, f_n(v))$ to every simple game v, where $f_i(v)$ represents the payoff to Player i and can be interpreted as a measure of the influence of Player i in game (N, v).

Absolute Banzhaf index β was introduced by Banzhaf [7] for any (N, v) and $i \in N$ as follows:

$$\beta_i(v) = \frac{|C_i|}{2^{n-1}},$$

where $C_i = \{S \subseteq N : i \in S, v(S) = 1 \text{ and } v(S \setminus \{i\}) = 0\}$. For a given Player i, $|C_i|$ denotes the number of winning coalitions in which this player is critical. This means his withdrawal changes a coalition from a winning one to a losing one.

Normalized Banzhaf index β' is defined as follows:

$$\beta'_i(v) = \frac{|C_i|}{\sum_{j \in N} |C_j|}.$$

More on this index and its property can be found in [12], for example. For the reason that this index takes a player's criticality in the winning coalitions into account, we chose β' to measure the power influence of a player in the European Parliament.

3 Some Data and Considerations on EU Before and After Brexit

Now, pursuant to Article 14(2) TEU of the Treaty of Lisbon, the 28-member-state EU counts 751 MEPs, which is the maximal number of representatives of the union's citizens. The minimum number of seats per member state is 6 (for example, Cyprus), while the maximum is 96 (Germany). Table 1 shows the current distribution of seats in the EP (see, the columns labeled "2014") and also lists the codes for each EU country provided in ISO 3166. After the withdrawal of the UK from the EU, this distribution of seats will change (see, the columns labeled "2019"). In February 2018, the EP decided to decrease the number of MEPs from 751 to 705 after Brexit comes into force. In this scenario, some EU countries will receive some additional MEPs, but no EU country will lose any seats (see Table 1 or [13]). The columns labeled "Corr." in Table 1 shows the planned corrections of seats for each member country of the EU due of the democracy changes of the last years.

The current and future proposed distributions of seats in the EP only take into account the population of the EU countries and follows the principle of degressive

Table 1. Country codes as well as current and future plan distribution of seats in EU Parliament.

Country	2014	2019	Corr.	Country	2014	2019	Corr.
Belgium (BE)	21	21		Lithuania (LT)	11	11	
Bulgaria (BG)	17	17		Luxembourg (LU)	6	6	
Czech Republic (CZ)	21	21		Hungary (HU)	21	21	
Denmark (DK)	13	14	+1	Malta (MT)	6	6	
Germany (DE)	96	96		Netherlands (NL)	26	29	+3
Estonia (EE)	6	7	+1	Austria (AT)	18	19	+1
Ireland (IE)	11	13	+2	Poland (PL)	51	52	+1
Greece (GR)	21	21		Portugal (PT)	21	21	
Spain (ES)	54	59	+5	Romania (RO)	32	33	+1
France (FR)	74	79	+5	Slovenia (SI)	8	8	
Croatia (HR)	11	12	+1	Slovakia (SK)	13	14	+1
Italy (IT)	73	76	+3	Finland (FI)	13	14	+1
Cyprus (CY)	6	6		Sweden (SE)	20	21	+1
Latvia (LV)	8	8		United Kingdom (GB)	73	0	

proportionality. So, members from lesser-populated countries have a relatively stronger presence in the EP.

Taking into account the decision-making in the EU, the Council together with the EP is generally the main decision-making body. Depending on the matter discussed, the Council takes its decisions by the application of votes: a simple majority (15 countries vote in favor), or the so-called weighted qualified majority (55% of the member states representing at least 65% of the EU population vote in favor), or unanimous vote (all member states vote in favor); see [14], for example.

For the distributions of seats in the EP from 2014 and 2019 (see Table 1), we calculated the distributions of power using the Banzhaf power index and simple majority of the MEPs; the result of these calculations are presented in Table 2. In the column labeled "Power ratio," we calculated the ratio of the Banzhaf index for 2019 to the Banzhaf index calculated for 2014. Hence, we can observe that all countries see an increase in their power in the EP after Brexit. Moreover, we do not observe that Brexit benefits the power of the greater-populated countries more than the lesser-populated countries in this case. But interestingly, increase in power after Brexit seems to have some correlation with a country's correction of seats in the EP – each of fourteen countries with amended number of MEPs made it into the list of top fourteen countries with largest power change coefficient (see Tables 1 and 2).

Table 2. Current and future distributions of seats and power in EU Parliament.

	2014				2019	
Seats	Banzhaf index	Country	Seats	Banzhaf index	Power ratio
18	0.0233	AT	19	0.0261	1.1202
21	0.0272	BE	21	0.0289	1.0625
17	0.022	BG	17	0.0234	1.0636
11	0.0142	HR	12	0.0165	1.162
6	0.0078	CY	6	0.0082	1.0513
21	0.0272	CZ	21	0.0289	1.0625
13	0.0168	DK	14	0.0192	1.1429
6	0.0078	EE	7	0.0096	1.2308
13	0.0168	FI	14	0.0192	1.1429
74	0.1002	FR	79	0.1152	1.1497
96	0.1366	DE	96	0.1465	1.0725
21	0.0272	GR	21	0.0289	1.0625
21	0.0272	HU	21	0.0289	1.0625
11	0.0142	IE	13	0.0178	1.2535
73	0.0987	IT	76	0.1103	1.1175
8	0.0103	LV	8	0.011	1.068
11	0.0142	LT	11	0.0151	1.0634
6	0.0078	LU	6	0.0082	1.0513
6	0.0078	MT	6	0.0082	1.0513
26	0.0338	NL	29	0.0399	1.1805

(*continued*)

Table 2. (*continued*)

Seats	Banzhaf index	Country	Seats	Banzhaf index	Power ratio
2014			2019		
51	0.0671	PL	52	0.0733	1.0924
21	0.0272	PT	21	0.0289	1.0625
32	0.0417	RO	33	0.0455	1.0911
13	0.0168	SK	14	0.0192	1.1429
8	0.0103	SI	8	0.011	1.068
54	0.0713	ES	59	0.0834	1.1697
20	0.0259	SE	21	0.0289	1.1158
73	0.0987	GB	0		

Regarding the member states' individual shares in the financing of the EU budget, we can say that each country's contribution (in the form of their own resources) consists of three parts (see [15], for example):

- customs tariffs and sugar levies collected on behalf of the EU (the so-called "traditional own resources"),
- GNI-based contribution, which is calculated as a fixed percentage of the gross national income (GNI),
- VAT-contribution calculated, which is calculated as a percentage of VAT income.

The GNI-based contribution is the largest part of each country's payment. As the GDP helps show the strength of a country's local income and the GNI helps show the economic strength of the citizens of a country, this was one of the motivations for us to modify the model proposed in 2005 and use the members states' contributions along with their populations as characteristics that could influence the assignment of seats in the EP. Table 3 shows member states' individual share in the financing of the EU budget in 2016. The list has been compiled using data published in the European Commission's financial report for 2016 (see [16]). The column labeled "Population" shows the number of persons having their usual residence in an EU country on 1 January 2016 (see [17]).

Table 3. Population and contributions of member states to EU budget in 2016.

Country	Population	EU contribution per country (in billions of euro)
Austria	8,700,471	2.763
Belgium	11,311,117	3.611
Bulgaria	7,153,784	0.382
Cyprus	848,319	0.152
Croatia	4,190,669	0.391
Denmark	5,707,251	2.207

(*continued*)

Table 3. (*continued*)

Country	Population	EU contribution per country (in billions of euro)
Estonia	1,315,944	0.183
Finland	5,487,308	1.829
France	66,730,453	19.476
Germany	82,175,684	23.274
Greece	10,783,748	1.509
Ireland	4,726,286	1.675
Italy	60,665,551	13.940
Latvia	1,968,957	0.218
Lithuania	2,888,558	0.319
Luxembourg	576,249	0.311
Malta	450,415	0.081
Netherlands	16,979,120	4.343
Poland	37,967,209	3.553
Portugal	10,341,330	1.587
United Kingdom	65,382,556	12.760
Czech Republic	10,553,843	1.361
Romania	19,760,314	1.374
Slovakia	5,426,252	0.646
Slovenia	2,064,188	0.34
Spain	46,440,099	9.564
Sweden	9,851,017	3.312
Hungary	9,830,485	0.924
Total	510,277,177	112.085

4   A Seat Apportionment Method

In this paper, we propose a model for apportioning seats in the EP. This method is a modification of the method we proposed in 2005 (see [1]). Let us remind you about the method. The general idea of the model from 2005 was not only to take into consideration one characteristic – the population of each country in assigning seats like it is in EU – but to also take each country's GDP into account. Since then, the governments of some countries (Italy, for example) have complained about their high contributions to the budget of the EU and their low representations in the EP; therefore, we decided to change the model proposed in [1] and take the total annual EU Contribution (CON) of each member country into consideration instead of its GDP. The new method consists of adequately weighing these two factors using a convex linear combination:

$$S_i = k \cdot P_i + (1 - k) \cdot C_i, \tag{1}$$

where P_i denotes the population of the i-th country (in percentages), C_i – the EU contribution of the i-th country to the EU budget (in %), k ($0 \leq k \leq 1$) – the population weighting of the i-th country, and S_i – the number of seats of the i-th country (in %). For instance, let the population and CON percentages of the i-th country be shown by P_i and C_i, respectively. Let us assume the weight for the population is 20% and the CON is 80%; in this case, seat percentages S_i of the i-th EU country will be $S_i = 0.2 \cdot P_i + 0.8 \cdot C_i$. To transform the seat percentages into the number of actual seats, a suitable rounding method can be used (e.g., d'Hondt's proportional system, Hamilton's Greatest Divisors, or the minimax apportionment proposed by Gambarelli in [18]).

The interesting question refers to the optimal value of k either for a particular country or for the whole EP. From our initial examinations, it seems that countries with high GNIs and, thus, high EU contributions prefer a lower value of k (preferably $k = 0$). However, the power influence of the member states in the EU calculated by the Banzhaf index, for example, is not a linear function of the numbers of seats. This fact was profoundly explained in [1] – also with a simple example. In the next section, we will underline this type of "paradox" by interpreting the results of the calculations made for the seat and power distributions in the EP using the data from 2016.

5   Application of Proposed Method

We applied the method proposed in the previous section to the EP, taking the latest available data of the 2016 member states' EU contributions from the official EU website (published in the European Commission's financial report for 2016 [16]) and the populations of the EU countries from Eurostat [17] (also 2016). All of this data is reported in Table 3. Because we are dealing with large numbers, we have converted the data into percentages and multiplied each by 100 for clarity before applying our new model. For example, the population of Italy was 60,665,551, which accounted for approximately 11.89% of the total population of the 28-country EU (see Table 3). Thus, we assigned the number 1189 for Italy in the column labeled "Only POP" ($k = 1$) in Table 4. In Table 6, we have assigned the number 1364 to Italy in the "Only POP" ($k = 1$) column since the total population of the EU will decrease to 444,894,621 after the withdrawal of the UK from the EU. Italy's population will then account for approximately 13.64% of the 27-country EU.

Table 4. Seat distribution in EP dependent on weight k (before Brexit).

	Only CON				Values of k						Only POP
	0.0	0.1	0.2	0.3	0.4	0.5	0.6	0.7	0.8	0.9	1.0
AT	247	239	231	224	216	209	201	193	186	178	170
BE	322	312	302	292	282	272	262	252	242	232	222
BG	34	45	55	66	76	87	98	108	119	130	140
HR	35	40	44	49	54	58	63	68	73	77	82
CY	14	14	14	14	15	15	15	16	16	16	17
CZ	121	130	139	147	156	164	173	181	190	198	207
DK	197	188	180	171	163	154	146	137	129	120	112
EE	16	17	18	19	20	21	22	23	24	25	26

(continued)

Table 4. (*continued*)

	Only CON				Values of k						Only POP
	0.0	0.1	0.2	0.3	0.4	0.5	0.6	0.7	0.8	0.9	1.0
FI	163	158	152	147	141	135	130	124	119	113	107
FR	1738	1695	1652	1609	1566	1523	1480	1437	1394	1351	1308
DE	2076	2030	1983	1937	1890	1843	1797	1750	1704	1657	1610
GR	135	142	150	158	165	173	181	188	196	204	211
HU	82	93	105	116	126	138	149	160	170	182	193
IE	149	144	138	132	127	121	115	110	104	98	93
IT	1244	1238	1233	1227	1222	1216	1211	1205	1200	1194	1189
LV	19	21	23	25	27	29	31	33	35	37	39
LT	28	31	34	37	40	43	45	48	51	54	57
LU	28	26	24	23	21	20	18	16	14	13	11
MT	7	7	8	8	8	8	8	8	8	9	9
NL	388	382	377	371	366	360	355	349	344	338	333
PL	317	360	402	445	488	531	573	616	659	701	744
PT	142	148	154	160	166	172	178	184	190	197	203
RO	123	149	176	202	228	255	281	308	334	361	387
SK	58	63	67	72	77	82	87	92	96	102	106
SI	30	31	32	33	34	35	36	38	38	39	40
ES	853	859	865	870	876	882	887	893	899	904	910
SE	296	285	275	265	254	244	234	224	213	203	193
GB	1138	1153	1167	1181	1196	1210	1224	1239	1253	1267	1281
Total	10,000	10,000	10,000	10,000	10,000	10,000	10,000	10,000	10,000	10,000	10,000

Based on our model (Eq. (1) in particular), we calculated the seat distribution in the EP to vary in weight k from 0 to 1 (using steps of one decimal place). More precisely, Table 4 shows the seat distribution for the 28-country EU (before Brexit), and Table 6 presents the seat distribution for the 27-country EU (after the proposed withdrawal of the UK from the EU).

For each country, the maximal percentage of seats assigned by our model are underscored in Table 4. For example, the maximal number of seats for Poland is obtained for $k = 1$ ($0.0744 * 751 \approx 56$ seats), while the minimal for $k = 0$ is ($0.0317 * 751 \approx 24$ seats). From an initial examination Table 4, it seems that the interests of countries with higher CON percentages than population percentages (France, Germany, Italy, etc.) are to have lower values of k (preferably 0), as the respective segments decrease. Conversely, for countries with lower CON percentages (Bulgaria, Poland, Romania, etc.), it seems that higher values of k (preferably 1) favor these countries in their numbers of seats, as the respective segments increase. However, this rule does not always apply (as can be illustrated in the case of Spain).

Taking into account the distributions of power presented in Table 5, we can see that the highest power for Spain is obtained for $k = 0.5$ or 0.6. However, the maximal number of seats ($0.0910 * 751 \approx 68$) is for $k = 1$ in Table 4. For $k = 0.5$, the percentage of seats is equal to 882, which is approximately equivalent to $0.0882 * 751 \approx 66$ seats. Therefore, Spain could obtain a more powerful position in the EP with a lower number of seats. Of course, if $k = 0.5$, it would be stated for the apportionment of seats in the EP. So, for the 28-country EU, the optimal values of k are whose indicated by the underscored values in Table 5; these represent the values that indicate the maximal power of each country in the EP. What is a bit surprising is that the optimal value of k is 1 for the UK. This means that it is better to calculate only the population of the UK for its seat apportionment.

Table 5. Power distribution in EP depending on weight k (before Brexit).

	Only CON				Values of k						Only POP
	0.0	0.1	0.2	0.3	0.4	0.5	0.6	0.7	0.8	0.9	1.0
AT	232	225	217	211	203	196	189	182	177	171	164
BE	302	294	284	275	265	255	246	238	230	223	216
BG	32	42	52	62	71	82	92	102	113	125	136
HR	33	38	41	46	51	54	59	64	69	74	80
CY	13	13	13	13	14	14	14	15	15	15	16
CZ	114	122	131	138	147	154	163	171	181	190	201
DK	185	177	169	161	153	145	137	129	123	115	109
EE	15	16	17	18	19	20	21	22	23	24	25
FI	153	149	143	138	132	127	122	117	113	108	104
FR	1770	1726	1682	1638	1595	1553	1510	1468	1423	1377	1328
DE	2261	2205	2149	2096	2044	1994	1945	1894	1843	1786	1727
GR	127	134	141	149	155	162	170	178	186	196	205
HU	77	87	99	109	118	130	140	151	162	175	188
IE	140	136	130	124	119	114	108	104	99	94	90
IT	1227	1220	1216	1213	1212	1211	1210	1207	1203	1197	1191
LV	18	20	22	23	25	27	29	31	33	36	38
LT	26	29	32	35	38	40	42	45	48	52	55
LU	26	25	23	22	20	19	17	15	13	13	11
MT	7	7	7	8	8	7	8	8	8	9	9
NL	362	358	354	349	344	338	334	329	327	325	325
PL	298	338	378	418	457	495	532	568	606	643	683
PT	134	139	145	150	156	161	167	174	181	189	197
RO	116	140	166	190	214	239	264	291	318	348	379
SK	55	59	63	68	72	77	82	87	91	98	103
SI	28	29	30	31	32	33	34	36	36	37	39
ES	875	887	899	907	915	920	920	918	913	905	897
SE	278	268	259	249	239	229	220	211	203	195	188
GB	1096	1117	1138	1159	1182	1204	1225	1245	1263	1280	1296
Total	10,000	10,000	10,000	10,000	10,000	10,000	10,000	10,000	10,000	10,000	10,000

Comparing the numbers in Table 4 with those in Table 6, we infer that no country loses seats after Brexit. This conclusion is obvious if the total numbers of seats in the EP will not change after Brexit. However, it is also true in case of a reduction in the

overall number of seats to 705. The last conclusion can be deduce after simple calculations. The values underscored in Table 6 indicate the maximal numbers of the
percentages of the seats obtained for a particular weighting k for each of the 27
countries in the EU (minus the UK). In order to determine the optimal values of k for
each country, it is necessary to reference Table 7.

Table 6. Seat distribution in EP depending on weight k (after Brexit).

	Only CON					Values of k					Only POP
	0.0	0.1	0.2	0.3	0.4	0.5	0.6	0.7	0.8	0.9	1.0
AT	278	270	262	253	245	237	229	220	212	204	196
BE	364	353	342	331	320	309	298	287	276	265	254
BG	39	51	63	75	87	100	112	124	136	148	161
HR	39	45	50	56	61	67	72	78	83	89	94
CY	15	16	16	16	17	17	18	18	18	19	19
CZ	137	147	157	167	177	187	197	207	217	227	237
DK	222	213	203	194	185	175	166	156	147	138	128
EE	18	20	21	22	23	24	25	26	27	28	30
FI	184	178	172	166	160	154	148	142	136	129	123
FR	1961	1915	1869	1823	1776	1730	1684	1638	1592	1546	1500
DE	2343	2294	2244	2194	2145	2095	2046	1996	1946	1897	1847
GR	152	161	170	179	188	197	206	215	224	233	242
HU	93	106	119	131	144	157	170	183	195	208	221
IE	169	162	156	150	144	137	131	125	119	112	106
IT	1404	1399	1395	1391	1387	1384	1380	1376	1372	1368	1364
LV	22	24	26	29	31	33	35	38	40	42	44
LT	32	35	39	42	45	49	52	55	58	62	65
LU	31	29	28	26	24	22	20	18	17	15	13
MT	8	8	9	9	9	9	9	9	10	10	10
NL	437	432	426	421	415	410	404	398	393	387	382
PL	358	407	457	506	556	606	655	705	754	804	853
PT	160	167	174	182	189	196	203	211	218	225	233
RO	138	169	199	230	261	291	322	352	383	414	444
SK	65	71	76	82	88	94	99	105	111	116	122
SI	34	35	37	38	39	40	42	43	44	45	46
ES	963	971	979	987	995	1003	1011	1020	1028	1036	1044
SE	334	322	311	300	289	277	266	255	244	233	222
Total	10,000	10,000	10,000	10,000	10,000	10,000	10,000	10,000	10,000	10,000	10,000

After Brexit, the optimal values of k are changed for some countries (e.g., Spain, Malta, and Slovenia). Now, the optimal value is $k = 1$ for Spain. For Malta, $k = 0.8$ and $k = 0.9$ are optimal, and for Slovenia – $k = 0.9$ and $k = 1$. For the rest of the countries, the optimal values of k are equal to those indicated by the underscored values in Table 7.

Table 7. Power distribution in EP depending on weight k (after Brexit).

	Only CON					Values of k					Only POP
	0.0	0.1	0.2	0.3	0.4	0.5	0.6	0.7	0.8	0.9	1.0
AT	260	253	247	240	233	227	219	211	202	194	184
BE	341	332	323	314	305	296	286	275	264	252	239
BG	36	48	59	71	83	95	107	119	130	140	152
HR	36	42	47	53	58	64	69	75	79	84	89
CY	14	15	15	15	16	16	17	17	17	18	18
CZ	128	138	148	158	168	179	189	198	207	215	223
DK	207	200	191	184	176	167	159	149	140	131	120
EE	17	19	20	21	22	23	24	25	26	27	28
FI	172	167	162	157	152	147	142	136	130	122	116
FR	1945	1913	1879	1842	1800	1758	1713	1667	1621	1576	1533
DE	2628	2579	2525	2465	2403	2335	2267	2197	2130	2066	2005
GR	142	151	160	170	179	188	197	206	214	221	228
HU	87	100	112	124	137	150	163	175	186	197	208
IE	158	152	147	142	137	131	125	120	113	106	100
IT	1537	1508	1481	1455	1432	1413	1398	1387	1381	1380	1382
LV	21	22	24	27	29	31	33	36	38	40	41
LT	30	33	37	40	43	47	50	53	55	59	61
LU	29	27	26	25	23	21	19	17	16	14	12
MT	7	7	9	8	9	9	9	9	10	10	9
NL	413	409	404	401	397	394	389	383	377	368	360
PL	336	384	435	485	538	592	647	704	760	818	874
PT	149	157	164	172	180	187	194	202	208	213	219
RO	129	158	188	218	249	279	309	338	367	394	418
SK	60	66	71	78	84	90	95	100	106	110	115
SI	32	33	35	36	37	38	40	41	42	43	43
ES	773	784	797	814	835	858	885	916	948	981	1014
SE	313	303	294	285	275	265	255	244	233	221	209
Total	10,000	10,000	10,000	10,000	10,000	10,000	10,000	10,000	10,000	10,000	10,000

In order to analyze change in power for each member state of the 27-country EU before and after Brexit, we calculated a power change coefficient for each country and each k in Table 8. This coefficient is calculate as follows:

$$\frac{\beta_i(v_{27}^k)}{\beta_i(v_{28}^k)},$$

where v_{27}^k denotes a 27-country weighted majority game with a simple majority and $\beta_i(v_{27}^k)$ is the Banzhaf index for country i calculated in Table 7 for a seat distribution given for a particular k after Brexit (see Table 6). Similarly, v_{28}^k denotes a 28-country weighted majority game with a simple majority and $\beta_i(v_{28}^k)$ is the Banzhaf index for country i calculated in Table 5 for a seat distribution given for a particular k before the withdrawal of the UK (see Table 4).

Analyzing Table 8, we can deduce that power increases for all countries except Spain after the UK's exit from the EU. We see that the power change coefficient is lower than 1 for Spain and for $k < 0.8$. So, only taking the EU contribution in the assignments of seats into account (i.e., $k = 0$) brings the worst situation for Spain after Brexit. This means that Spain loses power/influence in the 27-country EP for values of k up to 0.7. When we only take population into account, Spain and Poland take part in the group of the five greatest-populated countries in the EU (omitting just the UK). For both countries after Brexit, the optimal value of k is 1; however, its increase in power after Brexit is lower than for Poland even though Spain has a larger population than Poland. For Poland, the power ratio is 1.2796, and for Spain – only 1.1304. The largest increase in power is noted for Malta and $k = 0.5$. Continuing to only consider population ($k = 1$), we can conclude that Brexit benefits countries with large populations more than the others. The five countries with the largest populations are also the top five with the largest power change coefficients. This is similar to what was observed by the other researchers mentioned in Sect. 1 (see [4, 10, 11]). One explanation for this fact can be that, after Brexit, the positions of the four most-populated countries strengthens considerably. Namely, the top four countries in the 28-country EU (DE, FR, IT, ES) have at least 51.43% of the total number of seats in total (which is obtained for $k = 1$). After Brexit, these four countries will have at least 59.34% of the total number of seats in the 27-country EU (which is also obtained for $k = 1$).

Table 8. Power change coefficient (after Brexit).

	Only CON					Values of k					Only POP
	0.0	0.1	0.2	0.3	0.4	0.5	0.6	0.7	0.8	0.9	1.0
AT	1.1207	1.1244	1.1382	1.1374	1.1478	1.1582	1.1587	1.1593	1.1412	1.1345	1.122
BE	1.1291	1.1293	1.1373	1.1418	1.1509	1.1608	1.1626	1.1555	1.1478	1.13	1.1065
BG	1.125	1.1429	1.1346	1.1452	1.169	1.1585	1.163	1.1667	1.1504	1.12	1.1176
HR	1.0909	1.1053	1.1463	1.1522	1.1373	1.1852	1.1695	1.1719	1.1449	1.1351	1.1125
CY	1.0769	1.1538	1.1538	1.1538	1.1429	1.1429	1.2143	1.1333	1.1333	1.2	1.125
CZ	1.1228	1.1311	1.1298	1.1449	1.1429	1.1623	1.1595	1.1579	1.1436	1.1316	1.1095
DK	1.1189	1.1299	1.1302	1.1429	1.1503	1.1517	1.1606	1.155	1.1382	1.1391	1.1009
EE	1.1333	1.1875	1.1765	1.1667	1.1579	1.15	1.1429	1.1364	1.1304	1.125	1.12
FI	1.1242	1.1208	1.1329	1.1377	1.1515	1.1575	1.1639	1.1624	1.1504	1.1296	1.1154
FR	1.0989	1.1083	1.1171	1.1245	1.1285	1.132	1.1344	1.1356	1.1391	1.1445	1.1544
DE	1.1623	1.1696	1.175	1.1761	1.1756	1.171	1.1656	1.16	1.1557	1.1568	1.161
GR	1.1181	1.1269	1.1348	1.1409	1.1548	1.1605	1.1588	1.1573	1.1505	1.1276	1.1122

(continued)

Table 8. (*continued*)

	Only CON					Values of k					Only POP
	0.0	0.1	0.2	0.3	0.4	0.5	0.6	0.7	0.8	0.9	1.0
HU	1.1299	1.1494	1.1313	1.1376	1.161	1.1538	1.1643	1.1589	1.1481	1.1257	1.1064
IE	1.1286	1.1176	1.1308	1.1452	1.1513	1.1491	1.1574	1.1538	1.1414	1.1277	1.1111
IT	1.2526	1.2361	1.2179	1.1995	1.1815	1.1668	1.1554	1.1491	1.148	1.1529	1.1604
LV	1.1667	1.1	1.0909	1.1739	1.16	1.1481	1.1379	1.1613	1.1515	1.1111	1.0789
LT	1.1538	1.1379	1.1563	1.1429	1.1316	1.175	1.1905	1.1778	1.1458	1.1346	1.1091
LU	1.1154	1.08	1.1304	1.1364	1.15	1.1053	1.1176	1.1333	1.2308	1.0769	1.0909
MT	1	1	1.2857	1	1.125	1.2857	1.125	1.125	1.25	1.1111	1
NL	1.1409	1.1425	1.1412	1.149	1.1541	1.1657	1.1647	1.1641	1.1529	1.1323	1.1077
PL	1.1275	1.1361	1.1508	1.1603	1.1772	1.196	1.2162	1.2394	1.2541	1.2722	1.2796
PT	1.1119	1.1295	1.131	1.1467	1.1538	1.1615	1.1617	1.1609	1.1492	1.127	1.1117
RO	1.1121	1.1286	1.1325	1.1474	1.1636	1.1674	1.1705	1.1615	1.1541	1.1322	1.1029
SK	1.0909	1.1186	1.127	1.1471	1.1667	1.1688	1.1585	1.1494	1.1648	1.1224	1.1165
SI	1.1429	1.1379	1.1667	1.1613	1.1563	1.1515	1.1765	1.1389	1.1667	1.1622	1.1026
ES	0.8834	0.8839	0.8865	0.8975	0.9126	0.9326	0.962	0.9978	1.0383	1.084	1.1304
SE	1.1259	1.1306	1.1351	1.1446	1.1506	1.1572	1.1591	1.1564	1.1478	1.1333	1.1117
CV	5.51%	5.47%	5.43%	5.08%	4.23%	4.62%	3.82%	3.21%	3.44%	3.12%	3.88%

In the last row of Table 8, we calculated a coefficient of variation (CV) for each k. The CV represents the quotient of the standard deviation to the mean multiplied by 100%; we used this statistic for comparing the degrees of variation in the power changes for the different distributions of power determined by weighting k. In our case, these values of CV are small; the maximal is 5.51% (for $k = 0$), and the minimal – 3.12% (for $k = 0.9$). Hence, the differences in power change coefficients varying k from 0 to 1 in 1/10 steps is small on the one hand; however, the ratio of power between the 27 countries pre- and post-Brexit is rather insignificant on the other.

6   Concluding Remarks

In this paper, we have proposed a new method for seat apportionment in the EP (without entering into "politically correct" or "politically incorrect" aspects), taking into account two characteristics of each member state: population and contribution of each country to the EU budget. This method is a modification of the previous one proposed by Bertini, Gambarelli, and Stach in [1]. The question is about the optimal value of the weighting (denoted in this paper by k) for these two characteristics in order to have an optimal number of seats that gives each country maximal voting power. In Sect. 5, we applied our method to the EP before and after Brexit based on the data provided by the EP website and the Eurostat in 2016. In this way, we obtain the distributions of seats and power in the EP varying k (i.e., the weight of the population in our model) in 1/10 steps. Having the distribution of power, we can find an optimal value of k for each member state of the EU. This optimal value indicates the number of seats that gives each country a maximal value in terms of power. This optimal value of

k could serve for negotiation in the EU to state the weighting of population in order to apportion seats that takes not only the population of a country but also its economic strength into account.

The method could be improved in the sense that we could search for an optimal continuous range of weighting k. In this paper, we use only discrete values of k. The optimal weight interval for the i-th country is variability interval k (which guarantees the i-th country the maximal power index) and with the equal power index (the maximum number of seats).

In the paper, we used the Banzhaf power index to measure the power of each country in the EP. Of course, our approach is not limited to only one power index; other power indices could be used, such as Holler's PGI index [19, 20]. Applying the Holler index could be interesting, as this index takes only minimal winning coalitions into account when measuring the power of a player. However, the PGI index does not satisfy the local monotonicity property (also called the dominance property) in weighted majority games. Informally, this property states that a player with a larger weight cannot receive voting power that is lower than a player with a lesser weight (see [21], for example).

One further development could be the application of different power indices like sub-coalitional values (see, e.g., [22]) or a priori union indices (see, e.g., [23, 24]). The application of this kind of power measurement seems to be prominent, as the formation of groups of countries that present the same political interest (for example, sub-coalitions around the most populated countries, unions of Eurozone countries, or the alliance of the Central European Visegrád Group, and so on) is noted in some voting situations in the EP.

In the EP, the most used voting procedure is the so-called weighted qualified majority. Here, we ignored this procedure and used the simply majority of seats in order to calculate the power distribution in the EP.

Comparing the result with other authors, we see that we obtain a similar result if we only take one characteristic of the EU member states into consideration (i.e., population). Namely, the exit of the UK results in an increase of power of the largest-populated countries more than the others.

In [25–27], Słomczyński and Życzkowski proposed the so-called Jagiellonian Compromise method to assign voting weights to each EU country. The Jagiellonian Compromise approach is based on the Penrose square-root law [28, 29] as well as the appropriate choice of the optimal majority quota in the EU voting system. According to the Jagiellonian Compromise method, the voting weight attributed to each EU country is proportional to the square root of its population; the decision of the voting body is taken if the sum of the weights of the members of a coalition exceeds the majority quota given by the following formula:

$$q = \frac{1}{2}\left(1 + \frac{\sqrt{n_1} + \ldots + n_n}{\sqrt{n_1} + \ldots + \sqrt{n_n}}\right),$$

where n is the number of EU countries and n_i is the population of the i-th member state. Thus, we see that the value of q depends on the particular distribution of the population in the EU and n (the number of member states). This compromise solution may also be combined with a simple majority of states [30]. Note that the idea of assigning votes proportionally to the square root of a population is used in practice in the German Bundesrat to assign the number of representatives to each land. Moreover, some international organizations like EURAMET (The European Association of National Metrology Institutes), for example, use square-root systems to assign voting weights to its members proportional to the square root of each country's contributions to the budgets of the organizations. Then, it is known that the current voting system in the EU favors the most- and least-populated countries, and the Jagiellonian method restores some of the power to the medium-sized countries (from Spain to Ireland); see [27]. Thus, our method could also be modified in order to become a hybrid of a modification of the approach introduced by Bertini, Gambarelli, and Stach in [1] and the Jagiellonian method. In this way, we can obtain a more adequate nonlinear model that takes both the population and contribution factors into consideration and assigns weights to the EU countries proportionally to the square root of the weighted sum of both of these factors.

Acknowledgements. This research is financed by the funds (No. 16.16.200.396) of AGH University of Science and Technology, MIUR, research grants from the University of Bergamo, and the GNAMPA group of INDAM.

The authors would like to thank the anonymous reviewers for their numerous comments and suggestions that led us to improving this paper.

References

1. Bertini, C., Gambarelli, G., Stach, I.: Apportionment strategies for the European Parliament. In: Gambarelli, G., Holler, M. (eds.) Power Measures III, a Special Issue of Homo Oeconomicus, vol. 22, pp. 589–604. Accedo Verlagsgesellschaft, München (2005). Reprinted in: Holler, M.J., Nurmi, H. (eds.) Power, Voting, and Voting Power: 30 Years After, pp. 541–552. Springer-Verlag, Heidelberg (2013)
2. Bertini, C., Gambarelli, G., Stach, I.: A method of seat distribution in the European Parliament. In: Łebkowski, P. (ed.) Zarządzanie przedsiębiorstwem. Teoria i praktyka 2014, pp. 271–280. AGH University of Science and Technology Press, Kraków (2014)
3. Bilbao, J.M., Fernández, J.R., Jiménez Losada, A., López, J.J.: Generating functions for computing power indices efficiently. Top **8**(2), 191–213 (2000)
4. Mercik, J., Ramsey, D.M.: The effect of Brexit on the balance of power in the European Union Council: an approach based on pre-coalitions. In: Mercik, J. (ed.) Transactions on Computational Collective Intelligence XXVII. LNCS, vol. 10480, pp. 87–107. Springer, Cham (2017). https://doi.org/10.1007/978-3-319-70647-4_7
5. Shapley, L.S., Shubik, M.: A method for evaluating the distribution of power in a committee system. Am. Polit. Sci. Rev. **48**, 787–792 (1954)
6. Stach, I.: Shapley-Shubik index. In: Dowding, K. (ed.) Encyclopedia of Power, pp. 603–606. SAGE Publications, Los Angeles (2011)
7. Banzhaf, J.F.: Weighted voting doesn't work. A mathematical analysis. Rutgers Law Rev. **19**, 317–343 (1965)

8. Johnston, R.J.: On the measurement of power: some reactions to Laver. Environ. Plann. A **10**, 907–914 (1978)
9. Gładysz, B., Mercik, J., Ramsey, D.M.: The effect of Brexit on the balance of power in the European Union Council revisited: a fuzzy multicriteria attempt. In: Nguyen, N.T., Kowalczyk, R., Mercik, J., Motylska-Kuźma, A. (eds.) Transactions on Computational Collective Intelligence XXXI. LNCS, vol. 11290, pp. 80–91. Springer, Heidelberg (2018). https://doi.org/10.1007/978-3-662-58464-4_8
10. Kóczy, L.Á.: How Brexit affects European Union power distribution. Discussion Papers (MTDP-2016/11), Institute of Economics, Centre for Economic and Regional Studies, Hungarian Academy of Sciences, Budapest (2016)
11. Petróczy, D.G., Rogers, M.F., Kóczy, L.Á.: Brexit: the belated threat. arXiv:1808.05142v1 [econ.GN], 14 August 2018
12. Bertini, C., Stach, I.: Banzhaf voting power measure. In: Dowding, K. (ed.) Encyclopedia of Power, pp. 54–55. SAGE Publications, Los Angeles (2011)
13. EU elections: how many MEPs will each country get in 2019, News. European Parliament, EU affairs 01-02-2018 10:21 (2018). http://www.europarl.europa.eu/news/en/headlines/eu-affairs/20180126STO94114/eu-elections-how-many-meps-will-each-country-get-in-2019
14. Voting system. https://www.consilium.europa.eu/en/council-eu/voting-system. Accessed 3 Sept 2018
15. Keep, M.: A guide to the EU budget. Briefing Paper 06455 (2018). http://researchbriefings.files.parliament.uk/documents/SN06455/SN06455.pdf. Accessed 28 Oct 2018
16. The EU budget at a glance. http://www.europarl.europa.eu/external/html/budgetataglance/default_en.html. Accessed 3 Sept 2018
17. Eurostat, Code tps00001 - Population on 1 January – persons. https://ec.europa.eu/eurostat/tgm/table.do?tab=table&language=en&pcode=tps00001&tableSelection=1&footnotes=yes&labeling=labels&plugin=1. Accessed 12 Dec 2018
18. Gambarelli, G.: Minimax apportionments. Group Decis. Negot. **8**(6), 441–461 (1999)
19. Holler, M.J.: Forming coalitions and measuring voting power. Polit. Stud. **30**, 262–271 (1982)
20. Holler, M.J., Packel, E.W.: Power, luck and the right index. Zeitschrift für Nationalökonomie, J. Econ. **43**, 21–29 (1983)
21. Bertini, C., Freixas, J., Gambarelli, G., Stach, I.: Comparing power indices. Int. Game Theory Rev. **15**(2), 13400041–134000419 (2013)
22. Stach, I.: Sub-coalitional approach to values. In: Mercik, J. (ed.) Transactions on Computational Collective Intelligence XXVII. LNCS, vol. 10480, pp. 74–86. Springer, Cham (2017). https://doi.org/10.1007/978-3-319-70647-4_6
23. Owen, G.: Values of games with a priori unions. In: Henn, R., Moeschlin, O. (eds.) Essays in Mathematical Economics and Game Theory, vol. 141, pp. 76–88. Springer, Heidelberg (1977). https://doi.org/10.1007/978-3-642-45494-3_7
24. Owen, G.: Modification of the Banzhaf-Coleman index for games with a priori unions. In: Holler, M.J. (ed.) Power, Voting, and Voting Power, pp. 232–238. Physica, Würzburg (1982)
25. Życzkowski, K., Słomczyński, W.: Voting in the European Union: the square root system of Penrose and a critical point and optimal quota. arXiv:cond-mat/0405396 [cond-mat.other] (2004)
26. Słomczyński, W., Życzkowski, K.: Penrose voting system and optimal quota. Acta Physica Polonica **B37**(11), 3133–3143 (2006)

27. Słomczyński, W., Życzkowski, K.: Jagiellonian compromise – an alternative voting system for the council of the European Union. In: Życzkowski, K. (ed.) Institutional Design and Voting Power in the European Union, pp. 43–57. Routledge, London (2011)
28. Penrose, L.S.: The elementary statistics of majority voting. J. Roy. Stat. Soc. **109**(1), 53–57 (1946)
29. Penrose, L.S.: On the Objective Study of Crowd Behaviour. H.K. Lewis, London (1952)
30. Kirsch, W., Życzkowski, K., Słomczyński, W.: Getting the votes right. Eur. Voice **13**(17), 12 (2007)

The Use of Group Decision-Making to Improve the Monitoring of Air Quality

Cezary Orłowski[1]([✉]) [ID], Piotr Cofta[2] [ID], Mariusz Wąsik[1],
Piotr Welfler[1], and Józef Pastuszka[1]

[1] WSB University in Gdańsk, Gdańsk, Poland
corlowski@wsb.gda.pl
[2] University of Science and Technology (UTP), Bydgoszcz, Poland
piotr.cofta@utp.edu.pl

Abstract. The aim of this paper is to present the use of methods supporting group decision making for the construction of air quality measurement networks. The article presents a a case study of making group decisions related to the construction of a hybrid network for measuring air quality in Gdańsk. Two different methods of data processing were used in the decision making process. The first one is using fuzzy modeling for quantitative data processing to assess the quality of PM10 measurement data. The other is using trust metrics for the IoT nodes of four different measurement networks. The presented example shows the complexity of the decision making process itself as well as the choice of the method. The authors deliberately used both the quantitative and qualitative methods in the decision making process to show the need to search for the right method by decision-makers.

Keywords: Data quality · Decision-making · Trust management · Fuzzy logic

1 Introduction

Data collection of environmental data, such as the level of air pollution, through the civic IoT networks poses a challenge when it comes to data quality. Data is often collected on a voluntary basis, from measuring units of various quality, unevenly distributed throughout the area. The authors experienced this problem first-hand while constructing the air quality monitoring network in the area of Gdansk, Poland.

This problem may be approached through a combination of technological and procedural methods, such as over-sampling, self-cleaning, certification or regular service visits. Those methods however introduce a high cost that has not been acceptable within the scope of the project.

Instead, the authors explored the ability to distinguish between data of various quality through data post-processing, where data collected from various measuring units are cross-verified before being released for the use. This ability has been achieved within the project through a series of decisions. The focus of this paper is on the way group decision making has been used throughout the project to reach the objective of improving the data quality.

© Springer-Verlag GmbH Germany, part of Springer Nature 2019
N. T. Nguyen et al. (Eds.): TCCI XXXIV, LNCS 11890, pp. 127–145, 2019.
https://doi.org/10.1007/978-3-662-60555-4_9

The initial approach presented in this paper is based on the group of experts developing the model of data analysis that is based on fuzzy logic. This approach naturally invokes the notion of group decision making, as a group of experts had to agree on details of method (such as the fuzzy function).

The second approach was based on the use of methods derived from the trust management area, such as reputation-based systems or consensus-based systems to the problem at hand. Here, decision-making has been invoked twice: once to agree on the approach and details, but then also to delegate the actual decision-making to the automated system.

The paper is structured as follows. It started from the background information and the formulation of the research problem in the context of measurement of air quality. The brief introduction to social group decision-making is followed by the detailed presentation of the project used to collect data about air quality. Next, the solution based on expert cooperation and the use of fuzzy logic is presented and results are discussed. This is followed by the discussion of the possible use of other methods inspired by trust management. Conclusions close the paper.

2   Group Decision-Making

Group decision-making is a process where decisions are made by a group of people. However, some of its elements can be also used to facilitate the operation of the automated process, where group decision-making is used as a metaphor.

The group decision-making process is a complex one [16] in which at least two experts representing knowledge in a given field participate and determine their preference for making a joint decision [25, 26]. To assess this complexity, the challenge of group decision making has to be approached early in the project.

During the project described in this paper, the problem of group decision making became important. Decisions had to be made regarding the assessment of confidence in PM10 monitoring networks and their nodes, on the basis of a number of measurements, different in terms of value and quality of data, derived from different networks. For that, a technical solution had to be found. Hence experts were called to select the most appropriate solution.

As the question regarding the quality of data has been put forward to experts, two decision-making processes became visible: the human one, within the group of experts on what is the best approach to the problem and an automated one related to the processing itself. While the project started with the first approach, it eventually sought the automated solution, potentially supported by some human involvement, that mimics human group decision-making.

Due to the fact that a much larger group of experts took part in complex decision-making situations, the decision-making process requires the following steps:

- Selection of experts
- Evaluation of the complexity of the system which is assessed by experts
- Developing decisions using methods used by experts

Within the scope of this project, the project management sought a mixed group of experts from universities and the industry to work collaboratively on the problem.

In addition to expert knowledge, it is necessary to assure access to data relevant to the decision-making process. In the current world in which Big Data is widely available, one should consider how much the knowledge of experts involved in the decision process and its effects are actually derived from this knowledge of data. It should also be considered and to what extent the decision-making process depends on access to data and their size.

For this project, preliminary data gathered at the early stages of the network has been used by experts to ascertain the applicability of various methods.

Existing research highlights the importance of decision-making problems as well as the complexity of group decisions related to the uncertainty and knowledge of experts. Available studies on group decision problems are also the basis for the analysis of uncertainty in the case of monitoring nodes.

The following methods are used in group decision making: fuzzy modeling, preference analysis or classical semantic analysis. During the analysis, the decision making process in measurement networks was considered using the classical semantic analysis [16] both for the analysis of decision-making processes as well as decision modeling. For example, Wang and Hao [27] proposed using proportional language. Also [5, 9] presents a linguistic approach to making decisions.

It seems that the use of linguistic analysis is still the preferred path to the analysis of group decision-making processes. Therefore, on the one hand, classical methods of knowledge representation are used, such as instructions, rules, association rules, as well as methods of data mining [9, 27, 29]. It also includes the Bayes classification and grouping of data and grouping. For this reason, group decision making is a search area where you can also determine the suitability of linguistic processing methods [22]. It is also the direction of research, the results of which are presented in this work.

The use of case studies are important for decision-making, as it both informs the experts as well as indicate available solutions. For this reason, the presented article presents a case study of the construction of a measurement network and then shows examples of methods that can be (and were) used in the decision-making process. Therefore, this article should be treated as a case study showing the decision-making environment, the methods used and the decision-makers' solutions used in the decision-making process. For this reason, in the next part of the work presents the decision making environment of the project implemented in Gdańsk related to the construction of a hybrid quality measurement network.

After the installation of four monitoring networks in Gdańsk, the problem appeared how to assess the quality of the obtained data. The project had to deal with three low cost civic monitoring networks, generating low quality measurements off an opportunistically yet densely distributed set of units, and one automatic network with a small number of monitoring stations generating data of high quality.

The situation was further complicated by the fact that civic units were often left unattended for a prolonged period of time, leading to the further degradation of the quality of measurements. However, the objectives of the project could have been satisfied only with data from all the networks, so low quality nodes could not have been excluded.

In this situation, data mining methods had to be applied to data streams from monitoring units before releasing processed data for the use. The challenge of a decision-making process was thus to select the appropriate method or methods.

The project made certain assumptions about the approach to data mining that can be described as follows.

- The quality of data is a function of the operation of the monitoring unit. That is, at any point in time, the given unit produces data of certain quality. Such quality may change in time (both degrade and improve), but the unit itself submits no malicious results.
- The location of monitoring units can be both planned and unplanned, but the network itself is dense enough to provide more than one measurement from different units, at least for certain area.

Therefore it should be possible to determine the function that processes data from several sensors into a value that, with a given degree of accuracy, represents the actual value.

3   Motivation – the 'City Breathes' Project

The project that motivated this work was conducted in the IBM Advanced Research Center (IBM CAS) in Gdańsk, Poland. IBM CAS is a research environment located in the university. It integrates both specialists from IBM as well as university employees. In this environment, research projects are carried out for business partners as well as for the benefit of the host city.

The Center is able both to implement a web-based solution based on IoT nodes and, on the other hand, to produce software that supports the implementation of these works. For the city, CAS implements projects for the development of city management systems, according to the Smart Cities strategy. CAS employees also deal with the construction of mobile applications for IBM Rational products. They also deal with the construction of reference models supporting the software development cycle.

As a part of the research conducted by CAS, a project was launched with a public benefit organization such as NGO - FRAG (Gdańsk Agglomeration Development Forum) in Gdańsk under the name "Miasto Oddycha" (City Breathes) [7]. The aim of this project was to build a network of civic IoT nodes with the main purpose of the local monitoring of the PM10 dust concentration. A hybrid monitoring network consisting of several IoT networks and nodes was created, larger and more varied than the existing Armaag public station network.

The assumption of the project is that civic IoT nodes are to complement the existing monitoring network allowing for detailed measurement in selected locations. As the result, the total number of IoT nodes, including all networks, has doubled. During the network construction process, IoT nodes were made available to residents to encourage them to participate in the project.

The research project with NGO included the installation and implementation of four monitoring networks for the measurement of air pollution in Gdańsk by creating a Civic Measurement Network. The development of the network as well as the time of installation of IoT nodes was dependent on the level of involvement of residents in the process of installing measuring nodes on their properties. It was also contingent on the decision of FRAG.

In the first stage, measuring nodes purchased by one of the early business partners were used. Next, nodes manufactured as part of the student project at the School of Banking in Gdańsk were included in the monitoring network. Subsequently, existing measuring nodes of the City Hall were included. The next stage was the invitation to the Luftdaten project from Germany.

The selection of the Internet of Things nodes as well as the network development was coordinated by NGOs. While making decisions, the expansion of existing networks, adding new networks, improving the operation of individual IoT nodes as well as the expansion of automatic networks were taken into account. Both NGO representatives, city representatives and representatives of the university were involved in the decision making process.

Two different methods for the evaluation of existing measurement network nodes and extension possibilities were evaluated in the process of decision making, taking into account the quality of measurements obtained. It was found that the further development of the network will be possible based on the analysis of the quality of measurements. Next, the decision makers were given a solution allowing for the assessment of trust in the measurement stations that they could apply taking into account both the automatic stations which IoT nodes and measurement networks provided by individual partners.

4 The Civic Measurement Network

The Civic Measurement Network merges four monitoring networks: Armaag, WSB, Luftdaten and Airly. While sharing the same area of the city of Gdansk, they focused on different aspects of the monitoring, leading to nodes that are often incomparable in terms of cost, reliability, measurement process or quality of data.

4.1 The Armaag Network

The Armaag network [4] uses nine measurement stations for automatic measurement of PM10 dust and other substances: sulfur dioxide, nitric oxide as well as ammonia and benzene. The quality of data is considered high, in terms of reliability, veracity and accuracy. An example of the Armaag measurement node is shown on Fig. 1. The figure also shows the location of measurement stations and sample PM10 measurement results obtained in hourly cycles.

Analysis of the distribution of these measurement stations indicates that their number is inadequate to the needs of the area. Specifically, the question often arise about the level of PM10 dust in any place in the city, not just near the installed stations.

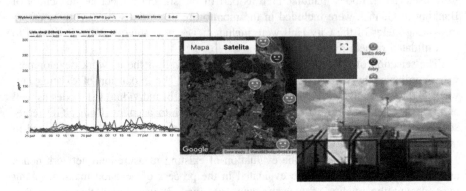

Fig. 1. The Armaag monitoring network

Significant costs of the Armaag measurement station limit the installation capacity of a larger number of such stations in Gdańsk. However, from the point of view of the quality of the civic network, the use of data from this small number of Armaag automatic stations can be crucial for the assessment of the quality of measurements at selected locations, as it can serve as a reference network, specifically if combined with pollution spreading model. This model can be used to locally estimate the concentration of dust in the air.

4.2   The WSB Network

Another monitoring network has been created as a part of a student project at WSB University. This network is built using IoT nodes that use Arduino Uno and Raspberry PI microcontrollers, selected for their ability to handle measurement sensors used to measure PM10 concentration. The process of building the node, the selection of software and the calibration of IoT nodes, created the conditions for the evaluation of the possibility of using this type of measurement stations in the construction of low-cost networks using low-cost measuring sensors.

The construction process (Fig. 2) also showed how complex it is to develop such a node. However, once created, the process of constructing of IoT nodes to measure PM10 is relatively simple and repeatable, allowing to produce nodes relatively fast.

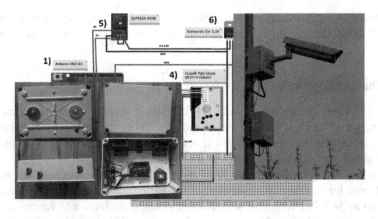

Fig. 2. Construction, IoT node and its deployment at the WSB in Gdansk

4.3   The Luftdaten Network

The Luftdaten project was created as part of the Open Knowledge Lab Stuttgart group of researchers and journalists programmers in Stuttgart [17]. Its implementation resulted from the analysis of significant air pollution in this city and the need to conduct continuous monitoring. The starting point for the construction of this network was the assumption that the sensor with the microcontroller should not cost more than 30 €. Eventually a network consisting of 300 independent IoT nodes measuring PM10 and PM2.5 pollutants in Stuttgart was created. Then the idea was transferred to other cities in Germany and Europe.

Fig. 3. IoT for measuring PM10 dust and a website with data presentation for Gdańsk

Currently, there are over 5600 such IoT nodes in Europe, of which 4,000 are installed in Germany. In Gdańsk, the installation of IoT PM10 and 2.5 nodes from Luftdaten was conducted through self-installation. During meetings with residents organized by the Civic Hub it was shown how the measurement of PM10 dust is carried out, but also the way of installing these IoT nodes. It turned out that due to the simplicity of the node, each of the residents after one meeting is able to install their own IoT node to measure PM10. Residents were also informed about the limitations associated with such a measurement method, such as low accuracy of measurement. Figure 3 shows the IOT node from Luftdaten and the site with measurement data for Gdańsk.

4.4 The Airly Network

The final monitoring network is Airly [1]. It is using small size IoT nodes similar to Luftdaten, aimed at local measurements and at a dense network. The installation of these measurement stations and their use was preceded by a thorough analysis of the need to install such measuring nodes in the city.

There are several reasons for those units to be installed. Because of their number, they allowed for the user of mathematical modelling that enabled predictions of air quality up to 24 h in advance. It also allowed to study the impact of various activities on the level of pollution. The network also increased residents' awareness of the pollution. The problem of air pollution in Poland is particularly important, as World Health Organization (WHO) standards are exceeded many times. Another reason is the issue of safety, i.e. indicating both places where the level of air pollution is low, as well as indicating what should be done to ensure safety for children, through the education of parents.

Fig. 4. IoT Airly nodes and the map of pollution for Gdańsk

It was also important that by creating a dense measuring network one could inform all those who are interested in outdoor activities like runners, cyclists and athletes in order to plan time of their activities so that the level of air pollution would be relatively low. Further, the network allows for dynamic response to changes, so that proper planning can take place. Figure 4 shows Airly stations and a map of pollution spreading.

5 Preliminary Data Analysis

Experts require data to make decisions, and in case of this project, they used measurements taken from the network, and applied simple forms of analysis to determine the nature of a problem. Their primary intention was to provisionally validate the assumption that only some stations degrade the overall quality of data. Experts worked with data similar to presented below.

Currently, the network consists of 23 stations of four different types. Some stations generate data at regular yet infrequent intervals of 30 or 60 min, some generate it less regularly, in about 5 min intervals. There is no guarantee that all readings always reach the server, so there are some missing readings. In total, the network generated more than 2,000 readings in any 24 h period.

The scatterplot of the recent readings of the PM 10 level over the relatively typical 24 h period is shown on Fig. 5. It is visible from the plot that, despite being geographically distributed, the majority of readings follow the similar pattern of gradual increases and decreases, defined by the general geography of the Gdansk area. However, there are some definitive outliers that cannot be easily explained.

If data cleansing [21] is applied to this kind of data set, it is likely to eliminate definitive outlier readings, i.e. probably those reporting the excessively high level of PM10. However, there may be problems with readings that gradually increase, specifically those towards the right side of the scatterplot, where it may be hard to discern between true outliers and local trends.

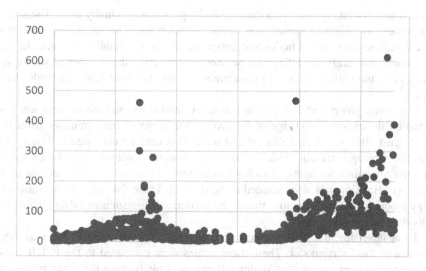

Fig. 5. Scatterplot of typical 24 h data points

Closer examination of selected stations (Fig. 6) shows that, against the typical backdrop demonstrated by station "1", only few stations such as "51" and "59" show unusual patterns of readings. That is, the majority of outliers and suspiciously-looking trends can be attributed to a small number of stations, thus validation one of the assumptions. If eliminated, or at least reduced in their impact on the final outcome, all remaining readings will provide information of higher quality. Thus the problem of data quality seems to lie with particular stations, not with individual readings.

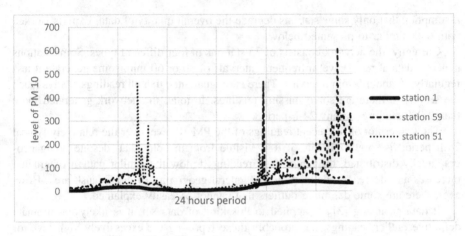

Fig. 6. Line plot for three stations, same time period

6 Fuzzy Logic Analyzis

In the decision-making process, a quantitative approach was initially used. The second assumption, as indicated earlier, was about the coverage. That is, the network must have a sufficient number of nodes to compensate for the unreliability of some. Experts determined that fuzzy modeling can be used to determine the number of nodes necessary to ensure higher quality of measurements. Only then trust-based methods can be used.

The initial group decision was to use fuzzy modelling, and the process was supported by the expert knowledge of the group. The range of measurements from July and August 2018 has been collected and used. This data set consisted of hourly data from the Armaag automatic stations and two measurement networks (Airly, Luftaden). Minute aggregation for hourly data from measurement stations built by students of the College of Banking was also included in the data set. These data allowed to construct a fuzzy model whose aim was to estimate the accuracy of measurement data obtained by individual IoT nodes.

To achieve this, the concentration with the highest accuracy of PM10 was determined with the fuzzy model. Then, these values were compared to the PM10 values obtained from the measurement stations. It was possible because the initial processing procedure was used using the clustering mechanisms. The application of these mechanisms created the conditions for estimating the data distribution and the use of standardization and normalization processes for data from the IoT nodes.

The principle was adopted that the choice of the final Data Mining method is a consequence of the preliminary data analysis. This preliminary data analysis presented in Fig. 7 shows the grouping at low, medium and high values from all networks.

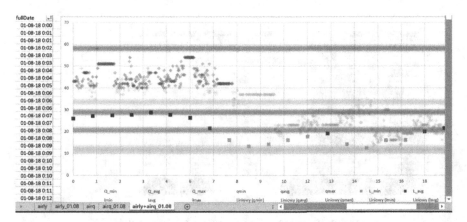

Fig. 7. Initial data analysis

The outcome of the preliminary data analysis indicated that the use of the fuzzy modelling can improve quality of the results from the network, thus satisfying one of the objectives of this step. Further, the use of fuzzy modelling was also considered to support the assessment of trust using data pre-processing based on clustering and evaluation of the quality of this data.

Such grouping of data creates conditions for building the functions of belonging and locating in the vertices of the membership functions in the centres of gravity of the pre-processed data. The centre of gravity method was used to place the apex of the membership function. Next, the fuzzy modelling procedure was applied, in which both the application process which took into account small average and high values based on expert knowledge was taken into account. The built-in rules were developed using the complete model construction procedures in which the number of rules was dependent on the number of input variables as well as on the granulation of these variables.

It was assumed that the number of input variables depends on the number of data obtained from four measurement stations. It was assumed that the value of the output variable is the exact value of the measurement. It was assumed that the station created with the use of a fuzzy model will allow to estimate the most accurate value. Figure 8 shows the results of measurements obtained with a fuzzy measurement station. It turns out that measurement errors are high both in the case of using sensor stations as well as automatic ones.

The analysis of measurement data coming from four measurement stations as well as automatic data stations obtained by means of a fuzzy model was the basis for defining trust in various elements of monitoring networks.

As the result, group decisions were made by the representatives of the Gdańsk Agglomeration Development Foundation regarding the possibility of expanding the existing monitoring network by additional monitoring stations. This decision was guided by the understanding of the quality of the data obtained. It stated that IoT nodes with a lower quality of measurement should form the basis for the further development of the network. The need to install additional monitoring stations was indicated, creating an even more extensive measurement network. Group decisions were also made

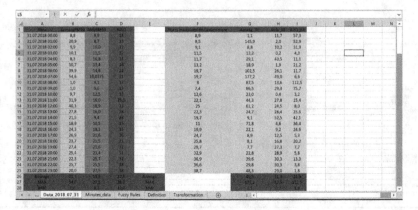

Fig. 8. Error values obtained from fuzzy modelling

in the construction of a fuzzy model in which the mechanism of inference was based on expert knowledge.

Further, with regard to the accuracy of measurements, decisions were also made when constructing and evaluating hourly data. Because hourly data was obtained on the basis of minute data and hence considered either taking into account the average measurement, the maximum value, the minimum value or the median. The choice of the average was based on the analysis of other hourly data and required expert assessments. In this expert assessment, other assessments were also taken regarding average values that were not published anywhere. It was assumed that all measuring networks with cheap measuring sensors acquire minute data and the hourly value is presented. However, none of the PM10 data providers presented the method under which the hourly value was determined on the basis of minute values.

The use of the fuzzy model used in the work allowed to estimate the accuracy of measurements of individual stations for the Gdańsk measurement networks. It also enabled the indication of those stations whose measurement quality is the lowest and indicated the necessity of extending the measurement networks by those whose measurement accuracy is high. It also enabled the support of the decision-making process regarding the expansion of the network with new IoT nodes.

7 Trust-Based Systems

As preconditions have been met, it was possible to introduce the notion of trust-based analysis to reduce the impact of ill-behaving nodes-stations. This was an important step from the perspective of a decision-making process. It is worth noting that up to now, the focus of the decision-making process was on a group of human, whether experts or decision-makers. Data itself was the subject of decisions, but parameters guiding those decisions were made by humans. Trust-based approach required experts to delegate at least some of their knowledge to the automated processing agents, something that was no apparent from the start [11].

Trust has been introduced here as a metaphor. That is, the experts did not assume that nodes are somehow 'trustworthy' in a human sense of this word, but merely observed that applying the simplified notion of trustworthiness to nodes can be beneficial for the overall network.

Trust-based approach required certain re-positioning of the problem. First, measurement data is no longer treated as a fact (whether true or not), but merely as an opinion of the node. Second, trustworthiness of the node must be automatically derived from data delivered by node, with no human intervention [14].

Research area of computational trust management is concerned with the collection, processing and the dissemination of trustworthiness and trust in its computational form (see [7] for an overview). It is inspired by social sciences that view trust as one of the major social enablers [10], but it has also its roots in information security where trust is seen as a foundation of applicable security measures [2]. Research such as Semantic Web [24] combined trust and provenance to determine the quality of information on the Web.

Trust and trustworthiness have several domain-specific definitions [18, 19], sometimes conflicting. If applied to the problem of data quality in in monitoring networks, one can define trust as the extent of rational reliance on monitoring data received from the unit. That is, the user of the data stream can trust data within the stream up to certain level, determined by trustworthiness of data. For example, more trustworthy data may be met with more trust in a form of important decisions being made while less trustworthy data will be met with less trust, for example with ignoring such data.

Note that trust management techniques can be incorporated into data processing of monitoring data in more than one way. While it may be expected to attribute trustworthiness to monitoring units, and use the provenance of data to determine trustworthiness of data, it is also possible to combine data of various levels of trustworthiness into a fuzzy set, or to post-process data with different level of trustworthiness into a more coherent (and more trustworthy) ones, in a manner know e.g. from the reputation-based systems.

Trust-based schemes can be classified depending on the source of trust that they can handle and it is useful to briefly describe such classification to determine which scheme can be applied to improve data quality, as discussed below.

7.1 Authority-Based Schemes

These are the very popular schemes, known primarily from information security [2]. Trustworthiness of a particular unit is determined here by the authority external to the scheme. For example, an administrator of a computer system may determine trustworthiness of its users and grant them different levels of access. Changes in trustworthiness must be again monitored by the external procedure (e.g. vetting) and corrections must be made manually.

If applied to the monitoring network, authority-bases scheme implies that trustworthiness of units should be determined by e.g. a group of experts. There is certain potential in such approach, as the experts already noticed that e.g. the Armaag network is expected to deliver more trustworthy results. However, authority-based scheme alone cannot provide flexibility and responsiveness when it comes to other networks.

7.2 Reputation-Based Schemes

The class of reputation-based schemes is both wide and popular [23], especially if one considers such schemes as Facebook's 'likes', eBay's reputation or Spotify recommendations. They all operate by automatically collating information about the extent of perceived trustworthiness from observers/consumers, centrally processing it into a reputation metric of producers/providers and distributing such reputation to interested parties, where it is accepted as the approximation of a trustworthiness of producers.

It is characteristic to those schemes that the trustworthiness is determined automatically by the system (without human intervention), automatically responds to changes in the behavior of producers (as reported by observers) and delivers an objective (or at least intra-subjective) understanding regarding the reputation.

Schemes do not value all opinions to the same extent, as not all opinions are always objective, well-intentioned and expressed. Thus schemes have to assess not only the trustworthiness of producers but also the trustworthiness of consumers, decreasing the impact of ill-behaving consumer on the overall trustworthiness. Schemes may also implement certain ageing of opinions, so that older opinions weight less towards the overall reputation. This allows to capture long-term changes in the behavior of the producer.

Reputation-based schemes, while inspired by social systems, found several applications in technology as well. All technological applications are underpinned by the same notion of providers delivering some services with varying degrees of trustworthiness and consumers, generating statements about individual experiences with providers, where consumers also have a varying degree of trustworthiness when it comes to the quality of their opinions. In some applications, the same technical component can play both roles: providers and customers.

Such approach is used e.g. in trust-based routing [15], where routers deliver service of different quality while components that wants to have their traffic routed can select the most trustworthy providers, while gossiping about their experience. Similarly, trust-based service composition [6] allows consumers to choose the service provider that they can trust on the basis of the experience of others. Cooperation among technical agents [28] uses trustworthiness to select the most trustworthy agents in open systems.

In the case of monitoring networks, such scheme should be able to calculate the reputation of the monitoring unit, on the basis of information submitted from other units. The main challenge lies in the fact that for the system to operate, there is a need to recreate the stream of opinions from what is currently only the stream of measurements. Let's consider a cluster two monitoring units that should produce similar results, but that deliver different ones. Assuming that no other information is available, it is unlikely to be possible to automatically determine which data is more trustworthy (i.e. which data more closely reflect the physical reality). Either unit may be trustworthy or not.

7.3 Evidence-Based Schemes

Evidence-based schemes operate by focusing on acquiring and retaining evidence about one's trustworthiness while deferring the determination of the level of such trustworthiness to interested parties (consumers). Each consumer can ten make their own decision regarding the trustworthiness of others on the basis of the available evidence. Contrasting with reputation-based schemes, there is no commonly shared notion of trustworthiness nor globally available reputations, so that decisions can be subjective.

Decision-making process, while left to consumers, can be quite complex (see e.g. [12]), generally mimicking rational human reasoning. The scheme itself is concerned only with the preservation of evidence. For that end, it uses technologies that increases confidence in data.

Blockchain technology (such as Bitcoin [20]) is the current example of such scheme at work. Nodes within the blockchain network manage the distributed database of evidence that is ever-growing, public and immutable. Individual nodes, however, have a duty to parse such database and resolve by themselves whether individual transactions.

Considering the challenge of data quality in the monitoring network, it is unlikely that the evidence-based scheme can be directly applied, as it is the objective of the project to achieve common understanding of trustworthiness. However, the fact that all measurement data (hence 'the evidence') will be stored in a database creates an opportunity to explore various algorithms to determine trustworthiness out of the common pool of evidence.

7.4 Introducing Trust into the Analysis of Data Quality

As trust is a social and psychological construct, it can be applied to data quality in monitoring networks only by analogy. As already mentioned, the key premise of the introduction of trust and trust-based reasoning will be the ability to automatically grade monitoring units and data they produce with certain level of trustworthiness. It does not imply that units have any kind of volition, but in a manner known from other trust-based solutions (e.g. [6]), trustworthiness will be attributed to nodes as a convenient metric. Such trustworthiness should be understood as the ability of the node to produce correct data.

Of three schemes presented earlier (authority-based, reputation-based and evidence-based), the architecture of the monitoring network lends itself to the reputation-based one. However, reputation-based systems flourish only if they are provided with the abundance of opinions that have comparable semantics, large base of responders, preferably separated from those who provide the service, of reasonable variety, and that are easy to process. Thus the main challenge for monitoring networks lies in extending the base of responders and introduce variety. For that, some elements of the authority-based scheme can be used.

It is therefore proposed here that there will be a way of introducing some initial variety into the level of trustworthiness of nodes, in the arbitrary way. Initially, this can be done manually (in a form of a bootstrap process), attributing initial higher trustworthiness to those nodes that are technically able to produce more correct results. For example, it has been already mentioned that one of the networks used in the experiment has nodes that produce results of higher quality.

This can be complemented by the (again, arbitrary and possibly expert-based) rule-based process that decrease trustworthiness of those nodes that produce apparently incorrect measurements, such as technically impossible values, physically impossible changes, unlikely drift etc. All those situations tend to indicate technical problems with the node, so that it is only fitting to use them to indicate the loss of trust in data produced by the node.

Apart from this bootstrapping activity and certain weeding off misbehaving nodes, nodes should be left to themselves to figure out which one is more trustworthy. Nodes do not produce separate opinions about other nodes, only a stream of measurement data. Therefore it is necessary to generate the stream of opinions about other nodes from such data stream.

Such generation may e.g. lead to opinions in a form of expectations that one node may have about measurements from other nodes, both past and future. Thus every node will not only produce results, but - indirectly - will also produce expectations regarding measurements that will be (or were) obtained by other nodes.

This task can be achieved only because nodes measure certain physical phenomena, such as the concentration of PM10 in the air. Those phenomena follow physical rules that make certain combination of measurements less believable than others. The already developed pollution spreading model incorporates this knowledge and allows to statistically interpolate the concentration of various pollutants from available measurements.

Due to the nature of both the phenomenon and the model, the confidence in such opinions decrease both in time and in space. Thus the confidence is higher when it comes to short-term predictions near the sensor, and becomes lower when the model has to predict the concentration ahead in time or far from the sensor.

Thus it is possible to determine two different aspects of trustworthiness of the node: one that is related to its ability to report and another that is related to the quality of its data. The former is the synthesis of the outcome of authority-based activities and the ability to make correct predictions, as dictated by the model. Note that, contrasting with social reputation-based systems, this trustworthiness is determined per relationship and per measurement. That is, the same node can have a different level of trustworthiness as a reporter in relation to different nodes, as those nodes can be at different distance. Further, the trustworthiness may vary depending on the temporal distance between the current and the predicted measurement.

Once the node can provide opinions about the trustworthiness of other nodes, it will be possible to introduce known algorithms to determine the trustworthiness of nodes when it comes to data quality.

7.5   The Use of Fuzzy Logic in Trust-Based Systems

As it has been discussed earlier, the project already used fuzzy modelling as a first approach, with good results. It is therefore useful to consider extending the use of fuzzy logic when it comes to trust based approach.

The use of fuzzy logic to determine the extent of trustworthiness has been already a research subject (e.g. [3, 13]). The key advantage is that fuzzy reasoning handles uncertainty and imprecision effectively, in a manner that can be easily comprehended by humans.

Fuzzy logic can be effectively used in recommendation-based systems to deliver the final value of trustworthiness, specifically when evidence of trustworthiness are incomparable or fuzzy by themselves. Further, it can be used to express a wide range of trust-related situations such as the lack of information or distrust (i.e. in nodes that seem to be overtaken by the adversary) [8].

Several aspects of the proposition presented here lend themselves to fuzzy values and fuzzy processing. For example, the level of confidence in the predictions provided by the model can be expressed in a form of fuzzy values and can be then a subject of processing according to the rules of fuzzy logic. In a similar manner, the reasoning about trustworthiness can be presented as fuzzy reasoning.

Consequently, it may be beneficial to implement fuzzy reasoning as a computational foundation of the trust-based approach.

8 Conclusion

This paper presents a case study of a process of decision-making that was used to resolve the challenge of data quality in heterogeneous monitoring networks that deal with air quality. Networks consist of a limited number of automatic stations with high measurement accuracy and a large number of low-cost measuring nodes with low measurement quality.

The approach assumed that data quality is a function of a monitoring unit that can be expressed as a trustworthiness of this unit. Following an expert-based group decision making, initial results were obtained through fuzzy modelling of both minute and hourly data, to determine trustworthiness of units.

Analysis of minute data indicated the need to use the median operator as the one that best corresponds to the value obtained from other measuring nodes. Group decisions were also made in the assessment of fuzzy rules built for the purpose of determining interpolated values. On the basis of this assessment, the trustworthiness of nodes was proposed as a solution to the problem of data quality. The accuracy of automatic station measurements was taken into account.

It seems that in the future the application of fuzzy modelling should take place both for the assessment of the location of measurement stations and for the assessment of measurement accuracy.

Complementing the solution could be the use of reputation-based schemes that are based on computational forms of trust. Such schemes were presented and discussed in this paper. If applied, the quality of data can be still expressed as trustworthiness of monitoring node, verified by measurements taken by other nodes, and consolidated using the physical model of the phenomena that are measured.

The authors expect that the combination of human-based group decision making and the automated processing controlled by computational trust algorithms may deliver significantly improved results.

References

1. Airly. https://airly.eu/pl/
2. Anderson, R.: Security Engineering: A Guide to Building Dependable Distributed Systems. Wiley, Hoboken (2001)
3. Aref, A., Tran, T.: A hybrid trust model using reinforcement learning and fuzzy logic. Comput. Intell. **34**, 515–541 (2018). https://doi.org/10.1111/coin.12155(2018)
4. Armaag. https://armaag.gda.pl/
5. Chabik, J., Orłowski, C., Sitek, T.: Intelligent knowledge-based model for IT support organization evolution. In: Szczerbicki, E., Nguyen, N.T. (eds.) Smart Information and Knowledge Management, pp. 177–196. Springer, Heidelberg (2010). https://doi.org/10.1007/978-3-642-04584-4_8
6. Chang, E., Dillon, T., Hussain, F.: Trust and Reputation for Service-Oriented Environments: Technologies for Building Business Intelligence and Consumer Confidence. Wiley, Hoboken (2006). ISBN: 978-0470015476
7. City Breathes. https://frag.org.pl/portfolio/trojmiastooddycha/
8. Cofta, P.: Trust, Complexity and Control: Confidence in a Convergent World, Wiley (2007). https://doi.org/10.1002/9780470517857. ISBN: 9780470061305
9. Dong, Y.C., Xu, Y., Yu, S.: Linguistic multiperson decision making based on the use of multiple preference relations. Fuzzy Sets Syst. **160**, 603–623 (2009)
10. Fukuyama, F.: Trust: The Social Virtues and the Creation of Prosperity, Touchstone Books (1996). ISBN: 0684825252
11. Golbeck, J.: Computing with Social Trust. Springer, London (2009). https://doi.org/10.1007/978-1-84800-356-9. ISBN: 978-1-84996-793-8
12. Górski, J., Cyra, Ł., Jarzębowicz, A., Miler, J.: Representing and appraising Toulmin model arguments in trust cases. In: Grasso, F. (et al.) The 8th International Workshop on Computational Models of Natural Argument (CMNA 8), Patras, Greece, 21 July 2008, 18th European Conference on Artificial Intelligence, 1 1, pp. 26–30 (2008)
13. Griffiths, N.: A fuzzy approach to reasoning with trust, distrust and insufficient trust. In: Klusch, M., Rovatsos, M., Payne, T.R. (eds.) CIA 2006. LNCS (LNAI), vol. 4149, pp. 360–374. Springer, Heidelberg (2006). https://doi.org/10.1007/11839354_26
14. Hewitt, E., Carpenter, J.: Cassandra: The Definitive Guide, 2nd edn. O'Reilly Media, Sebastopol (2016)
15. Jensen, Christian D., Connell, P.O.: Trust-based route selection in dynamic source routing. In: Stølen, K., Winsborough, W.H., Martinelli, F., Massacci, F. (eds.) iTrust 2006. LNCS, vol. 3986, pp. 150–163. Springer, Heidelberg (2006). https://doi.org/10.1007/11755593_12
16. Kacprzyk, J.: Group decision making with a fuzzy linguistic majority. Fuzzy Sets Syst. **18**, 105–118 (1986)
17. Luftdaten homepage: https://luftdaten.info/
18. McKnight, D.H., Chervany N.L.: The Meanings of Trust. In: University of Minnesota (1996). http://www.misrc.umn.edu/wpaper/wp96-04.htm
19. Nafi, K.W., Shekha kar, T., Hossain, A., Hashem, M.M.A.: An advanced certain trust model using fuzzy logic and probabilistic logic theory. Int. J. Adv. Comput. Sci. Appl. (IJACSA) **3**(12) (2012)
20. Nakamoto, S.: Bitcoin: A Peer-to-Peer Electronic Cash System (2008). https://bitcoin.org/bitcoin.pdf. Accessed 18 Dec 2018

21. Osborne, J.W.: Best Practices in Data Cleaning: A Complete Guide to Everything You Need to Do Before and After Collecting Your Data. Sage Publications (2012). ISBN: 978-1412988018
22. Pastuszak, J., Orłowski, C.: Model of Rules for IT Organization Evolution. In: Nguyen, N.T. (ed.) Transactions on Computational Collective Intelligence IX. LNCS, vol. 7770, pp. 55–78. Springer, Heidelberg (2013). https://doi.org/10.1007/978-3-642-36815-8_3
23. Resnick, P., Zeckhauser, R., Friedman, R., Kuwabara, K.: Reputation systems. Commun. ACM **43**(12), 45 (2000)
24. Richardson, M., Agrawal, R., Domingos, P.: Trust management for the semantic web. In: Fensel, D., Sycara, K., Mylopoulos, J. (eds.) ISWC 2003. LNCS, vol. 2870, pp. 351–368. Springer, Heidelberg (2003). https://doi.org/10.1007/978-3-540-39718-2_23
25. Rodríguez, R., Martınez, S., Herrera, F.: A group decision making model dealing with comparative linguistic expressions based on hesitant fuzzy linguistic term sets. Inf. Sci. **241**, 28–42 (2013)
26. Rodríguez, R.M., Martínez, L., Herrera, F.: Hesitant fuzzy linguistic term sets for decision making. IEEE Trans. Fuzzy Syst. **20**, 109–119 (2012)
27. Wang, J.H., Hao, J.: A new version of 2-tuple fuzzy linguistic representation model for computing with words. IEEE Trans. Fuzzy Syst. **14**(3), 435–445 (2006)
28. Wierzbicki, A.: Trust and Fairness in Open, Distributed Systems. Studies in Computational Intelligence, vol. 298. Springer, Heidelberg (2010). https://doi.org/10.1007/978-3-642-13451-7. ISBN: 978-3-642-13450-0
29. Wu, Z.B., Xu, J.P.: Possibility distribution-based approach for MAGDM With hesitant fuzzy linguistic information. IEEE Trans. Cybern. **46**, 694–705 (2016)

Bi-proportional Apportionments

Mirko Bezzi$^{(\boxtimes)}$ ⓘ, Gianfranco Gambarelli,
and Giuliana Angela Zibetti

Department of Management, Economics and Quantitative Methods,
University of Bergamo, Bergamo, Italy
{mirko.bezzi,gambarex,giuliana.zibetti}@unibg.it

Abstract. An apportionment method is proposed that generalises Hamilton's method for matrices, optimising proportionality in both directions, both for rows and columns. The resulting matrix respects fixed totals for rows and columns even when such totals do not satisfy standard criteria (monotonicity, maximum or minimum Hare), for example following the allocation of majority prizes to parties or coalitions.

Optionally, if required, the result can also respect the minimum Hare quotae for rows and columns. The algorithm may easily be expressed on the basis of rules.

Keywords: Bi-proportional · Apportionment · Electoral system · Calculation software

1 Introduction

This paper proposes an apportionment method that generalises that of Hamilton [5] for matrices, optimising proportionality in both directions, both for rows and columns. The resulting matrix respects fixed totals for rows and columns even when such totals do not satisfy standard criteria (monotonicity, maximum or minimum Hare): for example, due to the allocation of majority prizes to parties or to coalitions.

Optionally, if required, the result can also respect the minimum Hare quotae for rows and columns.

Over the following three sections, we deal with the problem of representativity and its applications, and we explain how this work may be applied to such contexts. The method proposed is described in Sects. 5 and 6, with certain characteristics of the solution presented in Sects. 7 and 8. An application to a recent case is given in Sect. 9. References for software are provided in the Appendix.

2 Representativity

Let us consider a population in which a subdivision is defined according to whether a component belongs to given economic, professional, biological or geographical categories, and so on. The problem of representation consists in associating such a

© Springer-Verlag GmbH Germany, part of Springer Nature 2019
N. T. Nguyen et al. (Eds.): TCCI XXXIV, LNCS 11890, pp. 146–161, 2019.
https://doi.org/10.1007/978-3-662-60555-4_10

population with a set with a lower cardinality (which may or may not be a subset of the first) that is able to describe it on the basis of established criteria. Other examples might be representatives in a board meeting who own shares in certain goods, or political party representatives in a Parliament related to votes received in elections, and so on.

3   Apportionments

In those cases in which the above-mentioned sets are described by integer vectors, it is usual to speak of apportionments. With a view to differing objectives, various apportionment criteria and methods have been studied; see, for example, the article by Gambarelli and Stach [4]. We limit ourselves here to mentioning the main ones, inasmuch as they are relevant to this paper. For simplicity of exposition, we shall refer to votes and seats, but what follows may equally well be applied to other contexts.

3.1   Objectives

There are two principal, although mutually opposed, goals in apportionment: representativity and governability. The former demands a distribution of seats as proportional as possible to that of the votes, in order to minimise the distance (using a suitable measure) between the percentages of votes and seats. Contrary to this, the latter demands a distribution of seats such as to guarantee a majority of seats for a party or preconstituted coalition. Given that these two goals are generally opposed, the tendency is to identify systems targeted at intermediate goals.

3.2   Criteria

Symmetry is a common criterion for both stated objectives. The apportionment must not depend on the order in which parties are considered when the apportionment method is implemented[1].

In terms of representativity and a majority prize for a party, common criteria are: the same number of seats for the same number of votes and monotonicity (not having fewer seats for majority votes). Such criteria are not valid in the instance of a majority prize for a coalition, since the relative parties can obtain seats as prizes, which gives them an advantage with respect to the others.

Respect of Hare quotae is among the criteria related to the sole objective of representativity. The Hare quota is defined as the quota of seats proportional to that of votes. Respect of the Hare minimum consists of the provision that seats assigned to each party will not be fewer than the Hare quota rounded down. Respect of the Hare maximum consists of the provision that seats assigned to each party will not be greater in number than the Hare quota rounded up. For brevity, we omit other criteria that have been proposed.

---

[1] From Gambarelli and Palestini [3].

At first sight, all of the preceding criteria would appear obvious, but, in given situations, some of them cannot be applied, for example if an odd number of seats must be assigned in a system made up of an even number of parties, where each party has received the same number of votes, it is impossible to respect the criterion of "the same number of seats for the same votes" and symmetry.

3.3   Methods

Apportionment methods biased towards partial or total governability use thresholds for parties with low voting percentages and/or techniques that favour the parties most voted for (including various types of large divisors), until the majority prize is awarded to a party, coalition, or relative majority.

One of the most well-known apportionment methods aimed at representativity is Alexander Hamilton's proportional system [5]. It consists in initially assigning seats equal to minimum Hare quotae, with a subsequent assignation of residual seats to parties with higher decimal places in their Hare quotae.

In all apportionment techniques, in the event of more than one distribution, a final choice is made on the basis of exogenous methods (i.e., in the case of elections, sex and/or age of candidates, the drawing of lots, and so forth).

4   Bi-apportionment

The problem of apportionment increases when the distribution must be made on bi-dimensional partitions, that is, the initial set is made up of votes obtained by various parties in various districts and the final set is made up of seats to be assigned to parties, with reference to pre-established totals for each district. In such cases, the problem is to transform a matrix of known integer elements (votes) into a matrix of unknown integer elements (seats), for which total seats have normally been given for each district (usually, in proportion to their population) and for each party (according to voting results and electoral regulations). Such problems are designated as bi-apportionment. For simplicity of exposition, in the course of this paper matrix rows will be called districts and columns will be called parties.

4.1   Infringement of Totals

As mentioned, district and party totals are given. Following this, seats are assigned within the matrix, bearing in mind both vectors of totals. Problems may arise from this process. Let us take, for instance, a matrix of votes and arrange the columns so that total votes for parties are in a non-decreasing order. Having determined total seats using a suitable apportionment method that respects monotonicity, there is a possibility that a distribution of seats within the matrix may not be found, such that it respects the monotonicity of all districts. Analogous problems may arise with regard to Hare quotae and other criteria.

4.2   Bi-apportionment in Multi-chamber Systems and Further Infringement of Totals

The problem of seat totals that fail to meet standard criteria may also arise from bi-apportionment in multi-chamber systems. Let us consider, for example, a two-chamber system in which there is a wish to award a national majority prize to the same party in both chambers to guarantee maximum governability. The party to be awarded the prize could be the one that has the largest total of votes related to both chambers. However, the party concerned may not have a relative majority (in terms of votes) in both chambers in which case the prize could infringe monotonicity at a national level, as well as other criteria, such as Hare quotae (maximum for the winning party, minimum for the others). This might happen even more in the case of national majority prizes involving a coalition. Analogous examples may be given for systems with more than two chambers.

The infringement of monotonicity and Hare quotae may also occur with regard to totals for districts, inasmuch as seats are assigned to the district on the basis of related population size, although the percentage of voters in a district may be different to the population of the district itself.

4.3   Bi-apportionment Methods

Having established totals for rows and columns, the problem remains of allocating seats within the matrix, respecting such totals. Various methods have been proposed to obtain seat matrices that are as proportional as possible (according to certain metrics) to those of votes. For further information on this, we refer the reader to the work by Demange [1]. However, such methods encounter difficulties, in sofar as for the most part they seek proportionality in a single direction, that is in regard to the rows, or to the columns, of the matrix. In such cases, they result in distorted effects, whose remedy, at times, may even involve a modification of the totals. By way of an example, in order to square the accounts during the Italian elections on February 24–25, 2013, an additional seat was assigned to Trentino-Alto-Adige and Sardegna (Sardinia), and one seat fewer to Friuli-Venezia-Giulia and Molise. Analogously, in previous legislation (2008), an additional seat was allocated to Veneto 1 and Piemonte (Piedmont) 2, and one seat fewer to Sicilia (Sicily) 1 and Trentino-Alto-Adige. In legislation preceding this (2006), an additional seat was assigned to Trentino-Alto-Adige and one seat fewer to Molise[2].

A method that respects line and column totals was introduced by Pukelsheim [7] (see also Pukelsheim et al. [8]). For an explanation of this method, see Sect. 9.

A general procedure was proposed by Gambarelli [2] and extended by Gambarelli and Palestini [3] to the bi-apportionment case. It involves a preliminary choice in the order of criteria to be respected (at a local level and/or at the level of totals). The process advances with progressively narrowing limits to the set of possible solutions, eliminating those that do not respect the first criterion, then those that remain that do

---

[2] For more information about calculation methods used for these elections, please refer to the law in force at the time of the elections [11].

not respect the second, and so on, skipping those criteria that would leave the set of remaining solutions empty. The disadvantage of this method is the computational complexity.

4.4 Our Intentions in This Paper

Our current proposal consists of an application of Gambarelli's and Palestini's procedure [3], prioritising criteria that determine totals and, following this, assigning seats at a local level, respecting, as the user prefers, first the minimum Hare quotae in the two directions and then the maximum proportionality in terms of Hamilton.

5 The Proposed Method

We shall now describe our method in simple terms.

5.1 Procedure

Once a table of total votes and seats to be assigned to each district has been determined (row totals), a preferred method is used to calculate the total numbers of seats to be assigned to parties (column totals).

A table of row Hare quotae is then created, each element of this being the product of votes obtained by a specific party in a specific district with regard to the total seats in that district, divided by the total votes in that district.

Analogously, a table is created for column Hare quotae, each element of which is the product of the votes obtained by the specific party in a specific district with regard to the total seats for the party, divided by the total votes for that party.

Then:

- if maximum preference is for the option of optimising proportionality, all table cells for seats under construction are zeroed;
- otherwise, each position is initially given a number of seats corresponding to the minimum between Hare quotae for row and column.

A reference matrix is then built, which, in each position, contains the maximum between Hare quotae for row and column.

After this, the following is applied:

LOOP.

- No further seats are awarded to districts and parties that have achieved relative totals.
- A seat is awarded in a position with the greatest difference between the element in the reference table and the number of seats assigned at present (all things being equal; see below).
- One seat is subtracted from the row and column total for seats still to be assigned, corresponding to that position.
- The loop cycle continues until there are no more seats to be assigned.

All things being equal, a comparison is made between the maximum differences obtained during the next step (and possibly in those following on from it) according to possible positions for a seat, and the position is chosen that shows the greatest difference. The exploration of the possible solutions does not influence the complexity of the implementation since, after a few steps, the same situation is always achieved.

The assignment of each seat does not preclude the allocation of subsequent seats because, at each step there is always one (or more) maximum values to which the next seat will be assigned. For this reason, the algorithm supplies one or more equivalent solutions.

6 An Example

Let us consider a parliament made up of three parties, A, B and C, and with two districts I and II, to which 4 and 6 seats are assigned respectively. Let us suppose that the votes received by the parties in an election are those given in Table 1 and that a national majority prize is awarded such as to give the party with a relative majority more than 50% of the seats, allocating the remaining seats to the other parties in proportion to the votes received, with numeric rounding following Hamilton. Seat totals are given in Table 2.

Table 1. Votes.

Parties/Districts	A	B	C	Totals
I	6	22	12	40
II	10	26	24	60
Totals	16	48	36	100

Table 2. Seat totals.

Parties/Districts	A	B	C	Totals
I				4
II				6
Totals	1	6	3	10

Our procedure begins with the calculation of line Hare quotae (Table 3). Therefore, C's Hare in the first district (=1.2) is obtained by dividing the votes received by C in that district (=12) by the total votes in that district (=40) and multiplying the result by the total number of seats in that district (=4).

Table 3. Line Hare quotae.

Parties/Districts	A	B	C	Totals
I	0.6	2.2	1.2	4
II	1.0	2.6	2.4	6

Column Hare quotae are calculated analogously (Table 4).

Table 4. Column Hare quotae.

Parties/Districts	A	B	C
I	0.38	2.75	1
II	0.62	3.25	2
Totals	1	6	3

At this point, we build a reference table (Table 5).

Table 5. Reference table.

Parties/Districts	A	B	C
I	0.6	2.75	1.2
II	1	3.25	2.4

Let us suppose that the option chosen is to favour respect of minimum Hare quotae. In this case, the matrix for seats is initialised as in Table 6.

Table 6. The initial matrix in the case of the option for favouring minimum Hare quotae.

Parties/Districts	A	B	C	To be added
I	0	2	1	1
II	0	2	2	2
To be added	1	2	0	3

In Table 7, we give the matrix of differences between the reference table (Table 5) and that for seats currently assigned (Table 6). Party C is removed from the count as all its seats have been allocated.

Table 7. The differences between Tables 5 and Table 6.

Parties/Districts	A	B	C
I	0.6	0.75	–
II	1	1.25	–

The largest element for such differences matrix is 1.25, which corresponds to party B in the second district. In this position, we therefore add a seat and update the number of seats still to be assigned. The result is given in Table 8.

Table 8. The matrix of seats provisionally assigned during the second step.

Parties/Districts	A	B	C	To be added
I	0	2	1	1
II	0	3	2	1
To be added	1	1	0	2

The new table of differences is given in Table 9.

Table 9. Differences related to the third step.

Parties/Districts	A	B	C
I	0.6	0.75	–
II	1	0.25	–

The next seat is assigned to party B in the first district because the largest element for differences matrix is 0.75. The last seat available is assigned to party A in the second district. Table 10 shows the final results.

Table 10. The solution in the case of a preference for minimum Hare quotae.

Parties/Districts	A	B	C
I	0	3	1
II	1	3	2

If, instead, the option had been to favour proportionality, the starting matrix would have contained only zeroes. In this case, too, acceptable results would have been those given below (Table 11).

Table 11. The solution in the case of a preference for the option of optimizing proportionality.

Parties/Districts	A	B	C
I	0	3	1
II	1	3	2

By way of contrast, in Table 12 we give the solutions we would have obtained using the Pukelsheim method adopted for elections in the Zurich District on February 12, 2006. Seat distribution coincides with the two solutions produced by means of the method we propose.

Table 12. The solution obtained with the Pukelsheim method (data calculated using BAZI software) [6].

Parties/Districts	A	B	C
I	0	3	1
II	1	3	2

7 Characteristics of the Solution

By construction, this method respects the following conditions:

- Monotonicity, defined by the assignment of priority to the position with greater Hare quota;
- Respect of row and column totals, due to the blocking of the assignment of seats;
- Minimum Hare quotae (row and column), since it uses them as a basis for initiating the assignation loop for remaining seats. Therefore, it is impossible to assign fewer seats than those corresponding to the minimum Hare quotae;
- Maximum Hare quotae (row and column), since the maximum number of seats that can be assigned corresponds to the maximum Hare.

The Pukelsheim method does not always guarantee respect of all these conditions (as we shall see in Sect. 9).

8 The Use of This Method in Italian Legislation

The method we propose assigns, at each step, a seat to the position in which there is the greatest difference between the reference table element and the number of seats currently assigned. A requirement for this system is that all parties are represented in all districts. In certain electoral systems, such as the Italian one, a party may choose not to be represented in all electoral districts, but only in some of them. In this case, to avoid seats being assigned in districts where the party is not represented, we have added a further control phase to the loop described in Sect. 5.1. This phase checks the remaining availability of seats to be assigned to each party only for those districts in which the party is a candidate. If this availability is equal to the number of total seats still to be assigned to the party at national level, the system assigns seats due to the given party directly, then it continues the assignation loop for seats for other parties/districts yet to be assigned.

9 A Comparison Between the New Method and Pukelsheim's Method

The laws adopted in the Italian electoral system in recent years have led to a non-respect of the row totals (*Mattarellum and Porcellum*) and those of the column (*Rosatellum*).

As has been shown in Sect. 4.3, this has resulted in district totals being adjusted to balance the figures. The procedure we propose, as with that proposed by Pukelsheim, does not lead to such distortions.

In Tables 13 and 14 we give votes and seats assigned during the elections on February 24–25, 2013, using the regulations in force. Complete data for the votes and seats assigned are to be found on the website for the Ministero dell'Interno "Archivio storico delle Elezioni" (Ministry of the Interior "Historical Archive of Elections"). Indicated in the same table ("diff") are the distortions introduced into district totals to guarantee that party totals balanced at a national level.

Table 13. Resulting votes to the Italian Chamber of Deputies during the political elections on February 24–25, 2013, following regulations in force [9].

Parties/Districts	CDE	PD	SEL	SVP	FDI	PDL	LN	SC	UDC	M5S	Total votes
Piemonte 1	3.787	358.768	49.562	0	26.839	237.410	43.966	139.753	13.976	393.079	1.267.140
Piemonte 2	2.790	285.095	26.624	0	39.091	269.174	78.400	130.870	16.763	313.573	1.162.380
Lombardia 1	4.373	638.627	68.974	0	35.074	476.981	200.214	258.452	20.414	472.154	2.175.263
Lombardia 2	5.037	580.837	47.076	0	37.335	518.705	442.669	274.783	31.985	462.797	2.401.224
Lombardia 3	2.901	248.016	19.056	0	17.493	196.392	98.120	78.271	11.764	191.195	863.208
Trentino A.A.	0	101.216	23.061	146.800	0	66.128	25.350	79.549	4.803	88.632	535.539
Veneto 1	3.388	363.768	29.962	0	29.948	344.649	194.033	178.631	29.683	458.082	1.632.144
Veneto 2	2.252	264.398	23.083	0	14.491	204.791	115.977	118.225	14.928	317.636	1.075.781
Friuli Ven.Giu.	2.346	178.001	17.880	0	12.920	134.118	48.310	77.557	11.633	196.037	678.802
Liguria	2.353	258.766	29.386	0	13.411	174.568	21.862	78.409	10.556	300.080	889.391
Emilia Roma.	6.062	989.810	77.312	0	35.990	434.534	69.108	211.777	29.568	658.475	2.512.636
Toscana	6.882	831.464	84.033	0	40.139	388.046	16.213	153.551	25.673	532.699	2.078.700
Umbria	1.512	168.726	16.772	0	14.563	102.329	3.081	41.366	6.796	142.959	498.104
Marche	3.572	256.886	27.744	0	19.993	162.480	6.405	78.210	16.737	298.114	870.141
Lazio 1	7.009	656.650	101.017	0	62.794	498.904	3.006	170.925	31.385	689.613	2.221.303
Lazio 2	2.514	196.186	26.762	0	28.750	257.799	2.869	53.660	18.425	240.880	827.845
Abruzzo	4.492	175.857	23.817	0	27.677	185.537	1.407	49.777	13.654	232.627	714.845
Molise	1.264	42.499	10.428	0	11.168	39.588	343	15.968	3.278	52.059	176.595
Campania 1	10.025	329.616	52.057	0	32.226	449.811	3.188	98.260	38.120	349.682	1.362.985
Campania 2	13.611	323.557	47.256	0	57.140	415.312	5.636	101.960	69.758	311.766	1.345.996
Puglia	32.054	407.279	144.465	0	34.264	637.815	1.578	172.307	45.567	562.398	2.037.727
Basilicata	8.009	79.631	18.357	0	7.397	59.171	382	24.569	7.960	75.260	280.736
Calabria	16.489	209.379	39.129	0	12.724	222.671	2.344	51.726	38.335	232.811	825.608
Sicilia 1	6.550	218.665	24.149	0	15.303	306.846	2.001	60.671	31.608	404.944	1.070.737
Sicilia 2	12.526	249.059	27.171	0	23.800	359.474	2.742	68.724	39.256	438.613	1.221.365
Sardegna	5.530	233.278	34.098	0	16.235	188.901	1.330	55.891	25.696	275.241	836.200
Totals votes	167.328	8.646.034	1.089.231	146.800	666.765	7.332.134	1.390.534	2.823.842	608.321	8.691.406	31.562.395

Table 14. Seats assigned to the Italian Chamber of Deputies during the political elections on February 24–25, 2013, following regulations in force.

Parties/Districts	CDE	PD	SEL	SVP	FDI	PDL	LN	SC	UDC	M5S	Total seats	Seats predicted	Diff.
Piemonte 1	0	11	2	0	0	3	1	2	0	4	23	23	–
Piemonte 2	0	10	1	0	1	3	1	2	0	4	22	22	–
Lombardia 1	0	21	2	0	1	5	2	3	0	6	40	40	–
Lombardia 2	0	20	2	0	0	7	6	4	0	6	45	45	–
Lombardia 3	0	8	1	0	1	2	1	1	0	2	16	16	–
Trentino A.A.	0	3	1	5	0	1	0	1	0	1	12	11	+1
Veneto 1	0	13	1	0	0	5	3	2	1	6	31	31	–
Veneto 2	0	9	1	0	0	2	2	2	0	4	20	20	–
Friuli Ven.Giu.	0	6	1	0	0	1	1	1	0	2	12	13	–1
Liguria	0	9	1	0	0	2	0	1	0	3	16	16	–
Emilia Roma.	0	28	2	0	0	5	1	2	0	7	45	45	–
Toscana	1	23	2	0	1	4	0	2	0	5	38	38	–
Umbria	0	5	0	0	0	1	0	1	0	2	9	9	–
Marche	0	9	1	0	0	2	0	1	0	3	16	16	–

(continued)

Table 14. (*continued*)

Parties/Districts	CDE	PD	SEL	SVP	FDI	PDL	LN	SC	UDC	M5S	Total seats	Seats predicted	Diff.
Lazio 1	0	21	3	0	1	6	0	2	1	8	42	42	–
Lazio 2	0	7	1	0	1	3	0	1	0	3	16	16	–
Abruzzo	0	6	1	0	0	3	0	1	0	3	14	14	–
Molise	0	2	0	0	0	0	0	0	0	0	2	3	–1
Campania 1	1	14	2	0	1	7	0	1	1	5	32	32	–
Campania 2	0	12	2	0	1	6	0	2	1	4	28	28	–
Puglia	1	15	5	0	1	9	0	2	1	8	42	42	–
Basilicata	0	3	1	0	0	1	0	0	0	1	6	6	–
Calabria	1	9	1	0	0	4	0	0	1	4	20	20	–
Sicilia 1	0	10	1	0	0	6	0	1	1	6	25	25	–
Sicilia 2	1	10	1	0	0	6	0	1	1	7	27	27	–
Sardegna	1	8	1	0	0	3	0	1	0	4	18	17	+1
Totals	6	292	37	5	9	97	18	37	8	108	617	617	–

In the following tables we give seat distribution as it would have been using the new method proposed (Table 15) and using the Pukelsheim method, adopted during elections in the Zurich District (Table 16), with a subsequent comparison of the two results obtained (Table 17).

Table 15. The results that would have been produced for the Italian Chamber of Deputies during the political elections on February 24–25, 2013, if the method proposed in this paper had been used instead of the regulations in force.

Parties/Districts	CDE	PD	SEL	SVP	FDI	PDL	LN	SC	UDC	M5S	Total seats	Seats predicted	Diff.
Piemonte 1	0	12	1	0	0	3	0	2	0	5	23	23	–
Piemonte 2	0	10	2	0	1	3	1	2	0	3	22	22	–
Lombardia 1	0	20	1	0	0	7	2	4	0	6	40	40	–
Lombardia 2	0	19	1	0	0	7	8	4	0	6	45	45	–
Lombardia 3	0	9	1	0	0	2	1	1	0	2	16	16	–
Trentino A.A.	0	3	1	5	0	0	0	1	0	1	11	11	–
Veneto 1	0	12	1	0	1	4	3	3	1	6	31	31	–
Veneto 2	0	10	1	0	0	2	2	2	0	3	20	20	–
Friuli Ven.Giu.	0	7	1	0	0	1	1	1	0	2	13	13	–
Liguria	0	9	1	0	0	2	0	1	0	3	16	16	–
Emilia Roma.	0	29	1	0	0	5	0	2	0	8	45	45	–
Toscana	0	24	1	0	0	5	0	2	0	6	38	38	–
Umbria	0	6	1	0	0	1	0	0	0	1	9	9	–
Marche	0	9	1	0	0	2	0	1	0	3	16	16	–
Lazio 1	0	21	2	0	0	7	0	2	0	10	42	42	–

(*continued*)

Table 15. (*continued*)

Parties/Districts	CDE	PD	SEL	SVP	FDI	PDL	LN	SC	UDC	M5S	Total seats	Seats predicted	Diff.
Lazio 2	0	8	2	0	1	3	0	0	0	2	16	16	–
Abruzzo	0	8	1	0	1	2	0	0	0	2	14	14	–
Molise	0	2	1	0	0	0	0	0	0	0	3	3	–
Campania 1	1	11	2	0	1	8	0	2	1	6	32	32	–
Campania 2	1	12	2	0	1	6	0	1	1	4	28	28	–
Puglia	1	13	4	0	0	11	0	3	1	9	42	42	–
Basilicata	1	4	1	0	0	0	0	0	0	0	6	6	–
Calabria	1	9	2	0	0	3	0	1	1	3	20	20	–
Sicilia 1	0	8	2	0	1	5	0	1	1	7	25	25	–
Sicilia 2	1	9	1	0	1	6	0	1	1	7	27	27	–
Sardegna	0	8	2	0	1	2	0	0	1	3	17	17	–
Totals	6	292	37	5	9	97	18	37	8	108	617	617	–

As mentioned above, Pukelsheim used a method quite similar to ours. According to this method, we start by calculating all the Hare quotae and the decision on the seat to be assigned is taken step by step, in the course of the process, and not on the basis of the maximum Hare quota (as in our case), but on the average of the row and column Hare quotae for each element.

Table 16. The results that would have been produced for the Italian Chamber of Deputies if, instead of the regulations in force, the Pukelsheim method had been employed, as adopted for the election in the Zurich District (data calculated using BAZI software) [6].

Parties/Districts	CDE	PD	SEL	SVP	FDI	PDL	LN	SC	UDC	M5S	Total seats	Seats predicted	Diff.
Piemonte 1	0	12	1	0	0	3	1	2	0	4	23	23	–
Piemonte 2	0	10	1	0	1	3	1	2	0	4	22	22	–
Lombardia 1	0	21	2	0	1	6	2	3	0	5	40	40	–
Lombardia 2	0	20	1	0	1	6	6	4	1	6	45	45	–
Lombardia 3	0	9	1	0	0	2	1	1	0	2	16	16	–
Trentino A.A.	0	2	1	5	0	1	0	1	0	1	11	11	–
Veneto 1	0	13	1	0	1	4	3	2	1	6	31	31	–
Veneto 2	0	9	1	0	0	2	2	2	0	4	20	20	–
Friuli Ven.Giu.	0	6	1	0	0	2	1	1	0	2	13	13	–
Liguria	0	9	1	0	0	2	0	1	0	3	16	16	–
Emilia Roma.	0	28	2	0	0	5	1	2	0	7	45	45	–
Toscana	0	24	2	0	1	4	0	2	0	5	38	38	–
Umbria	0	5	1	0	0	1	0	0	0	2	9	9	–
Marche	0	8	1	0	0	2	0	1	0	4	16	16	–
Lazio 1	0	22	3	0	1	6	0	2	0	8	42	42	–
Lazio 2	0	7	1	0	0	4	0	1	0	3	16	16	–

(*continued*)

Table 16. (*continued*)

Abruzzo	0	6	1	0	0	3	0	1	0	3	14	14	–
Molise	0	1	0	0	0	1	0	0	0	1	3	3	–
Campania 1	1	13	2	0	1	7	0	2	1	5	32	32	–
Campania 2	1	12	2	0	1	6	0	1	1	4	28	28	–
Puglia	2	15	5	0	1	9	0	2	1	7	42	42	–
Basilicata	0	3	1	0	0	1	0	0	0	1	6	6	–
Calabria	1	9	2	0	0	3	0	1	1	3	20	20	–
Sicilia 1	0	10	1	0	0	5	0	1	1	7	25	25	–
Sicilia 2	1	10	1	0	0	6	0	1	1	7	27	27	–
Sardegna	0	8	1	0	0	3	0	1	0	4	17	17	–
Totals	6	292	37	5	9	97	18	37	8	108	617	617	–

The advantage of our method lies in approaching the solution through a global MaxMin principle that can exclude extreme situations. Furthermore, our method has two possible variants, one of which guarantees respect of the minimum Hares unlike Pukelsheim's method.

Moreover, the algorithmic nature of Pukelsheim's method means they are difficult to translate into legislative terms, so that the Swiss Cantons that adopted them (such as the Canton in which Zurich is located) were obliged to refer directly to the software in related legislation.

Table 17. Differences between the two methods.

Parties/Districts	CDE	PD	SEL	SVP	FDI	PDL	LN	SC	UDC	M5S
Piemonte 1	0	0	0	0	0	0	−1	0	0	+1
Piemonte 2	0	0	+1	0	0	0	0	0	0	−1
Lombardia 1	0	−1	−1	0	−1	+1	0	+1	0	+1
Lombardia 2	0	−1	0	0	−1	+1	+2	0	−1	0
Lombardia 3	0	0	0	0	0	0	0	0	0	0
Trentino A.A.	0	+1	0	0	0	−1	0	0	0	0
Veneto 1	0	−1	0	0	0	0	0	+1	0	0
Veneto 2	0	+1	0	0	0	0	0	0	0	−1
Friuli Ven.Giu.	0	+1	0	0	0	−1	0	0	0	0
Liguria	0	0	0	0	0	0	0	0	0	0
Emilia Roma.	0	+1	−1	0	0	0	−1	0	0	+1
Toscana	0	0	−1	0	−1	+1	0	0	0	+1
Umbria	0	+1	0	0	0	0	0	0	0	−1
Marche	0	+1	0	0	0	0	0	0	0	−1
Lazio 1	0	−1	−1	0	−1	+1	0	0	0	+2
Lazio 2	0	+1	+1	0	+1	−1	0	−1	0	−1
Abruzzo	0	+2	0	0	+1	−1	0	−1	0	−1
Molise	0	+1	+1	0	0	−1	0	0	0	−1

(*continued*)

Table 17. (*continued*)

Parties/Districts	CDE	PD	SEL	SVP	FDI	PDL	LN	SC	UDC	M5S
Campania 1	0	−2	0	0	0	+1	0	0	0	+1
Campania 2	0	0	0	0	0	0	0	0	0	0
Puglia	−1	−2	−1	0	−1	+2	0	+1	0	+2
Basilicata	+1	+1	0	0	0	−1	0	0	0	−1
Calabria	0	0	0	0	0	0	0	0	0	0
Sicilia 1	0	−2	+1	0	+1	0	0	0	0	0
Sicilia 2	0	−1	0	0	+1	0	0	0	0	0
Sardegna	0	0	+1	0	+1	−1	0	−1	+1	−1

As may be seen from Tables 15 and 16, both solutions resolve distortions for the total of seats assigned to various districts, while Table 17 shows that differences in distribution between the two systems are significantly reduced (a maximum variation of two seats in a smaller number of cases).

However, if we carry out a check with regard to minimum Hare quotae, we discover that, while our system respects them in all distribution cells, the Pukelsheim method does not assign a sufficient number of seats to comply with the minimum Hare. In the example given of elections for the Italian Chamber of Deputies on February 24–25, 2013, there are 3 cases in which the minimum Hare is violated:

(1) The PDL in Toscana has a column Hare of 5.13[3]. Our method assigns 5 seats (equal to the truncated minimum Hare), while the Pukelsheim method assigns only 4 seats;
(2) The M5S in Emilia Romagna has a column Hare of 8.18[4]. Our method assigns 8 seats (equal to the truncated minimum Hare), while the Pukelsheim method assigns only 7 seats;
(3) The M5S in Toscana has a column Hare of 6.62[5]. Our method assigns 6 seats (equal to the truncated minimum Hare), while the Pukelsheim method assigns only 5 seats.

10 Changes Introduced by the Italian Electoral Law 2017

To solve the distortion problem of the total seats by districts, the new Italian electoral law (November 3, 2017) introduced a cross-district compensation, in order to reassign some seats where the previously analyzed errors occur. The use of this compensation,

---

[3] Hare 5.13 obtained by dividing the 388,046 district votes by the 7,332,134 national votes and multiplying by 97 total seats to be assigned to the party at a national level.

[4] Hare 8.18 obtained by dividing the 658,475 district votes by the 8,691,406 national votes and multiplying by 108 total seats to be assigned to the party at a national level.

[5] Hare 6.62 obtained by dividing the 532,699 district votes by the 8,691,406 national votes and multiplying by 108 total seats to be assigned to the party at a national level.

however, creates a new problem, that of non-compliance with monotonicity in the allocation of seats for each district.

The election result for the Chamber of Deputies of March 4, 2018, shows an example of non-compliance with monotonicity in the Molise district (Table 18), where the only available seat was assigned to the fourth ranked party instead of the first.

Table 18. Allocation of seats for the Italian Chamber of Deputies for the Molise district during the political elections on March 4, 2018 [10]

Party	Votes	Decimal part of the attribution quotient	Assigned Seats	Seat Compensation	Final Seats
Movimento 5 Stelle	78.079	0,472833	1	**−1**	
Forza Italia, Lega, Fratelli d'Italia, UDC	51.992	0,314854			
PD, +Europa, SVT-PATT, Civica Popolare, Italia Europa Insieme	28.568	0,173003			
Liberi e Uguali	6.491	0,039308		**1**	1
Totals	165.130				

Acknowledgments. This paper is under the patronage of MIUR. The authors wish to thank Luciano Violante for his valuable comments on a previous version of this paper, and Angelo Uristani for useful discussions on a local level seat assignation method.

Appendix A: Legislation

A draft legislative rule, referring to the case of the minimum Hare quotae, could be the following:

- Table of the row Hare quotae is determined, each element of which is the product of the votes obtained by that party in that district for the total seats in that district, divided by the total votes in that district;
- Table of column Hare quotae is determined, each element of which is the product of the votes obtained by that party in that district for the total seats of that party, divided by the total votes in that party;
- An *assigned seats matrix* is prepared, initially assigning the minimum value truncated, between the row Hare and the column Hare for each position;
- The *reference matrix* is then prepared, assigning the maximum value between the row Hare and the column Hare for each position;

- At each step, the method assigns a seat to the position in which there is the greatest difference between the *"reference matrix"* element and the *"assigned seats matrix"* element. The assigned seat is updated in the *"assigned seats matrix"*;
- The method checks the residual availability of seats that may be assigned to each party only for those seats for which a given party is represented and when residual availability is equal to the remaining number of seats to be assigned to the given party at a national level;
- The system continues in the assignment loop until all the seats have been allocated.

Appendix B: Automatic Calculation Software

The latest version of the bi-proportional apportionment software described in this paper is available at the following web address:
    http://dinamico2.unibg.it/dmsia/staff/gampubl.html#software.

References

1. Demange, G.: On allocating seats to parties and districts: apportionments. In: Fragnelli, V., Gambarelli, G. (eds.) Open Problems in Applications of Cooperative Games - a Special Issue of International Game Theory Review, vol. 15, no. 3 (2013)
2. Gambarelli, G.: Minimax apportionments. Group Decis. Negot. **8**(6), 441–461 (1999)
3. Gambarelli, G., Palestini, A.: Minimax multi-district apportionments. Homo Oecon. **24**(3/4), 335–356 (2007)
4. Gambarelli, G., Stach, I.: Power indices in politics; some results and open problems. Essays in honor of Hannu Nurmi. Homo Oecon. **26**(3/4), 417–441 (2009). ISBN 978–3-89265-072-0, ISSN 0943-0180
5. Hamilton, A.: Proposal of apportionment approved by U.S.A. Congress (1791)
6. Maier, S., Pukelsheim, F.: BAZI: a free computer program for proportional representation apportionment. Preprint Nr. 042-2007. Institut fur Mathematik, Universitat Augsburg (2007). www.opus-bayern.de/uni-augsburg/volltexte/2007/711/
7. Pukelsheim, F.: Current issues of apportionment methods. In: Simeone, B., Pukelsheim, F. (eds.) Mathematics and Democracy: Recent Advances in Voting Systems and Collective Choice. Studies in Choice and Welfare, pp. 167–176. Springer, Heidelberg (2006). https://doi.org/10.1007/3-540-35605-3_12
8. Pukelsheim, F., Ricca, F., Scozzari, A., Serafini, P., Simeone, B.: Network flow methods for electoral systems. Networks **59**, 73–88 (2011). Special Issue on the INOC 2009 Conference, April 26–29, 2009, Pisa, Italy
9. Website of the Italian Ministry of Interior "Historical Archives of the Elections", edited by the Office IV - Election Informatics Services. (http://elezionistorico.interno.it). Accessed 19 Dec 2015
10. Minutes of operations of the Central National Electoral Office of March 20, 2018 (http://www.cortedicassazione.it/corte-di-cassazione/it/elezioni.page). Accessed 26 Apr 2018
11. Italian Law December 21, 2005, n. 270 Changes to the rules for the election of the Chamber of Deputies and the Senate of the Republic (2005)

A Probabilistic Unified Approach for Power Indices in Simple Games

Josep Freixas[(⊠)] and Montserrat Pons

Department of Mathematics, Universitat Politècnica de Catalunya
(Campus Manresa), EPSEM, Avda. Bases de Manresa, 61-73, 08242 Manresa, Spain
{josep.freixas,montserrat.pons}@upc.edu

Abstract. Many power indices on simple games have been defined trying to measure, under different points of view, the "a priori" importance of a voter in a collective binary voting scenario. A unified probabilistic way to define some of these power indices is considered in this paper. We show that six well-known power indices are obtained under such a probabilistic approach. Moreover, some new power indices can naturally be obtained in this way.

Keywords: Simple games · Power indices · Probabilistic models

Math. Subj. Class. (2000): 91A12 · 91A40 · 91B12

1 Introduction

Different ways to assign a measure of power to voters in collective decision–making processes have been proposed and analyzed for many authors. The most well–known are the Banzhaf [3,5,13] (also known as the Penrose index) and the Shapley–Shubik [14] indices, but other indices as those defined by Johnston [12], Deegan and Packel [6], Alonso and Freixas [1] (also known as shift index), Holler [10] (also known as the Public Good index) or Alonso–Freixas–Molinero index [2] have also interesting properties. Several power indices for simple games are addressed to the problem of dividing among players a unit of a fixed divisible prize worth, say, 1 unit of transferable utility (TU) (e.g., money, a cake). These power indices, by their nature, are efficient: nothing of the divisible good is wasted. They also share the common fact that are based on some subsets of winning coalitions for which players in them are critical, i.e., after the deletion of a player in a winning coalition the coalition becomes losing.

In this context, the importance of a voter, seen as his/her estimated gain, can be measured as his/her expected value under a probabilistic model. Such a probabilistic model is determined by providing an answer for each one of the three following questions:

© Springer-Verlag GmbH Germany, part of Springer Nature 2019
N. T. Nguyen et al. (Eds.): TCCI XXXIV, LNCS 11890, pp. 162–170, 2019.
https://doi.org/10.1007/978-3-662-60555-4_11

- What kind of winning coalitions can be formed.
- What is the probability of each coalition to be formed.
- How the benefits are to be distributed among players in the formed coalition.

This point of view was explicitly used in [6] to define the Deegan–Packel index, by assuming that only minimal winning coalitions can be formed, that all minimal winning coalitions have the same probability to be formed and that the shares are divided equally among the members of the victorious coalition. Our contribution in this note is to show that many other power indices in simple games can be defined under the same probabilistic scheme by assuming different answers to the questions above. In fact, the crucial question is the second one because: (a) the more natural answer to the third question is to divide the benefits equally, and (b) certain coalitions are assumed to be impossible, i.e., the probability of their formation is zero, so that the answer to the second question can include the answer to the first one.

The rest of the paper is organized as follows. Section 2 is devoted to recall the fundamentals on simple games. In Sect. 3, two groups of three well-known power indices are seen under similar probabilistic approaches, following what we call Model 1 and Model 2. In Sect. 4 three new indices are introduced under the same approach under what we call Model 3. In Sect. 5 it is proved that the three indices based on winning coalitions containing crucial players are ordinally equivalent in complete simple games, result that complements the well-known equivalence ordinal between the Johnston and the Banzhaf indices on complete simple games. The Conclusion ends the paper in Sect. 6.

2   Basics on Simple Games

In the sequel, $N = \{1, 2, \ldots, n\}$ will denote a fixed but otherwise arbitrary finite set of *players*. Any subset $S \subseteq N$ is a *coalition*. A *cooperative game* is a map $v : 2^N \to \mathbb{R}$ such that $v(\emptyset) = 0$. A cooperative game v (in N, omitted hereafter) is a *simple game* if (a) $v(S) = 0$ or 1 for all S, (b) is monotonic, i.e. $v(S) \leq v(T)$ whenever $S \subset T$, and (c) $v(N) = 1$. Either the family of *winning* coalitions $W = W(v) = \{S \subseteq N : v(S) = 1\}$ or the subfamily of *minimal* winning coalitions $W^m = W^m(v) = \{S \in W : T \subset S \Rightarrow T \notin W\}$ determines the game. The players $a \in N$ that can convert a winning coalition into a losing one by leaving the coalition are called *critical* players:

a is critical in a coalition S if and only if $a \in S$, $S \in W$ and $S \setminus \{a\} \notin W$.

The set of *critical coalitions* is $W^c = W^c(v) = \{S \in W : S \setminus \{i\} \notin W,$ for some $i \in N\}$.

Observe that $W^m \subseteq W^c$.

A player $a \in N$ is *null* if $a \notin S$ for all $S \in W^m$. A player $a \in N$ has *veto* if $a \in S$ for all $S \in W$.

To define the last subset of winning coalitions we are interested in, we need to introduce the desirability relation, given first in [11]. Let v be a simple game on N and $a, b \in N$:

$$a \succsim b \quad \text{if and only if} \quad S \cup \{b\} \in \mathcal{W} \Rightarrow S \cup \{a\} \in \mathcal{W} \quad \text{for all} \quad S \subseteq N \backslash \{a, b\}.$$

It is not difficult to check that $\succsim$ is a preordering on N. $\succsim$ (resp., $\succ$) is called the *desirability* (resp., *strict desirability*) relation and $\approx$ the *equi–desirability* relation. The relationship between the desirability relation and power indices have been intensively studied, among them we refer to [4,7,8,15,16].

A simple game is *complete* if the the desirability relation is a complete preordering, i.e., either $a \succsim b$ or $b \succsim a$ for all pair of players a, b.

A coalition S is *shift-minimal* winning if $S \in \mathcal{W}$ and $(S \setminus \{a\}) \cup \{b\} \notin \mathcal{W}$ for all $a \in S$ and $b \notin S$ with $a \succ b$. The set $\mathcal{W}^s = \mathcal{W}^s(v)$ denote the set of shift-minimal winning coalitions and satisfies: $\mathcal{W}^s \subseteq \mathcal{W}^m \subseteq \mathcal{W}^c$. Only for $n \geq 4$ there are simple games in which the inclusion $\mathcal{W}^s \subseteq \mathcal{W}^m$ is strict.

Example 1. Let $(N, \mathcal{W}^m)$ be the simple game, with $N = \{1, 2, 3, 4, 5\}$ and

$$\mathcal{W}^m = \{S \subseteq N : \mid S \mid = 3 \quad and \quad S \neq \{3, 4, 5\}\}.$$

It holds $1 \approx 2 \succ 3 \approx 4 \approx 5$. Thus, the coalitions $\{1, 2, 3\}$, $\{1, 2, 4\}$ and $\{1, 2, 5\}$ are minimal winning but not shift-minimal winning because, for instance, $\{1, 3, 5\}$ is still winning and $2 \succ 3$.

Finally, let's establish some notation. The three collections of winning coalitions that are used to define power indices in the next section are either $\mathcal{W}^c$, $\mathcal{W}^m$ or $\mathcal{W}^s$. $\mathcal{Y}$ refers to anyone of these collections when generic properties or notations are to be established.

- For each $a \in N$, $\mathcal{Y}_a$ denote the set of coalitions in $\mathcal{Y}$ containing a.
- $\mathcal{C}_a(\mathcal{Y}) = \{S \in \mathcal{Y}_a : S \backslash \{a\} \notin \mathcal{W}\}$ i.e., $\mathcal{C}_a(\mathcal{Y})$ denote the set of coalitions in $\mathcal{Y}_a$ for which a is *critical*. Throughout the paper we denote its cardinality with $\eta_a(\mathcal{Y}) = |\mathcal{C}_a(\mathcal{Y})|$.
  Note that:

  - $\mathcal{C}_a(\mathcal{W}^m) = \mathcal{W}_a^m$, thus $\eta_a(\mathcal{W}^m) = |\mathcal{C}_a(\mathcal{W}^m)| = |\mathcal{W}_a^m|$.
  - $\mathcal{C}_a(\mathcal{W}^s) = \mathcal{W}_a^s$, thus $\eta_a(\mathcal{W}^s) = |\mathcal{C}_a(\mathcal{W}^s)| = |\mathcal{W}_a^s|$.
  - $\mathcal{C}_a(\mathcal{W}^c) \subseteq \mathcal{W}_a^c$, thus $\eta_a(\mathcal{W}^c) = |\mathcal{C}_a(\mathcal{W}^c)| \leq |\mathcal{W}_a^c|$.

But the inclusion $\mathcal{C}_a(\mathcal{W}^c) \subseteq \mathcal{W}_a^c$ can be strict. For instance, in Example 1 note that

$$\mathcal{W}_3^c = \{\{1, 2, 3\}, \{1, 3, 4\}, \{1, 3, 5\}, \{2, 3, 4\}, \{2, 3, 5\}, \{1, 2, 3, 4\}, \{1, 2, 3, 5\}\}$$

$$\mathcal{C}_3(\mathcal{W}^c) = \{\{1, 2, 3\}, \{1, 3, 4\}, \{1, 3, 5\}, \{2, 3, 4\}, \{2, 3, 5\}\}.$$

- For $S \in \mathcal{W}^c$:

  $\mathcal{C}(S) = \{i \in S : S\backslash\{i\} \notin \mathcal{W}\}$, i.e., $\mathcal{C}(S)$ is the set of critical players in S ($\mathcal{C}(S) \neq \emptyset$).

  $\chi(S) = \dfrac{1}{|\mathcal{C}(S)|}$, i.e., $\chi(S)$ is the converse of the number of critical players in S.

  Note that $\dfrac{1}{|S|} \leq \chi(S) \leq 1$, but if $S \in \mathcal{W}^m$ or $S \in \mathcal{W}^s$ then $\chi(S) = \dfrac{1}{|S|}$.

3   Known Power Indices for Simple Games

In the next subsections we consider two probabilistic models which lead to the definition of six already known indices. The only difference between the two models is the assumption on the probability of a coalition to be formed. In all cases $\mathcal{Y}$ indicates a collection of winning coalitions. Although the collections used in this note are either $\mathcal{Y} = \mathcal{W}^c$, $\mathcal{Y} = \mathcal{W}^m$ or $\mathcal{Y} = \mathcal{W}^s$, other possibilities could be also considered.

Let us remark that our approach is not necessarily the way in which every particular index was defined by their authors but it turns out to be true that all of them can be expressed under this unified scheme. The model 1 was already introduced in [6] for $\mathcal{Y} = \mathcal{W}^m$ but, as far as we know, model 2, which leads to three well–known power indices was never explicitly stated.

3.1   Model 1

This model is defined by the following assumptions:

- Only winning coalitions in $\mathcal{Y}$ will be formed.
- All coalitions in $\mathcal{Y}$ have equal probability of being formed.
- All critical members of the victorious coalition in $\mathcal{Y}$ receive equal shares of the 'spoils', while the rest of players in this coalition receive nothing.

Under these assumptions, the probability of a coalition $S \in \mathcal{Y}$ to be formed is $P(S) = \dfrac{1}{|\mathcal{Y}|}$ and the payoff of a player $a \in S$ if $a \in \mathcal{C}(S)$ is $\dfrac{1}{|\mathcal{C}(S)|} = \chi(S)$, while her payoff is 0 if $a \notin \mathcal{C}(S)$. Thus, the **Expected value of a's gain** if a is not null is:

$$E_a(\mathcal{Y}) = \sum_{S \in \mathcal{C}_a(\mathcal{Y})} \frac{1}{|\mathcal{Y}|}\chi(S) = \frac{1}{|\mathcal{Y}|} \sum_{S \in \mathcal{C}_a(\mathcal{Y})} \chi(S) \tag{1}$$

and $E_a(\mathcal{Y}) = 0$ if a is a null player.

For each collection $\mathcal{Y}$ of winning coalitions the formula (1) gives the value of a power index, obtaining in this way the three indices shown below. In the next definitions we assume that $a \in N$ is not a null player because in this case all indices are declared to be zero.

Johnston index $(\mathcal{Y} = \mathcal{W}^c)$

$$J_a = \frac{1}{|\mathcal{W}^c|} \sum_{S \in \mathcal{C}_a(\mathcal{W}^c)} \chi(S)$$

Deegan–Packel index $(\mathcal{Y} = \mathcal{W}^m)$

$$DP_a = \frac{1}{|\mathcal{W}^m|} \sum_{S \in \mathcal{C}_a(\mathcal{W}^m)} \chi(S) = \frac{1}{|\mathcal{W}^m|} \sum_{S \in \mathcal{W}_a^m} \frac{1}{|S|}$$

Alonso–Freixas–Molinero index $(\mathcal{Y} = \mathcal{W}^s)$

$$AFM_a = \frac{1}{|\mathcal{W}^s|} \sum_{S \in \mathcal{C}_a(\mathcal{W}^s)} \chi(S) = \frac{1}{|\mathcal{W}^s|} \sum_{S \in \mathcal{W}_a^s} \frac{1}{|S|}$$

3.2   Model 2

This model is defined by the following assumptions:

- Only winning coalitions in $\mathcal{Y}$ will be formed.
- The probability that a coalition in $\mathcal{Y}$ is formed is proportional to the number of critical players it contains.
- All critical members of the victorious coalition in $\mathcal{Y}$ receive equal shares of the 'spoils', while the rest of players in this coalition receive nothing.

Under these hypothesis, for each possible election of the set $\mathcal{Y}$, the probability of a coalition $S \in \mathcal{Y}$ to be formed is $P(S) = \dfrac{|\mathcal{C}(S)|}{\sum_{T \in \mathcal{Y}} |\mathcal{C}(T)|}$ and the payoff of a player $a \in S$ is the same as in the model 1. Thus, the **Expected value of a's gain** is:

$$E_a(\mathcal{Y}) = \sum_{S \in \mathcal{C}_a(\mathcal{Y})} \frac{|\mathcal{C}(S)|}{\sum_{T \in \mathcal{Y}} |\mathcal{C}(T)|} \chi(S) = \frac{|\mathcal{C}_a(\mathcal{Y})|}{\sum_{T \in \mathcal{Y}} |\mathcal{C}(T)|} = \frac{\eta_a(\mathcal{Y})}{\sum_{T \in \mathcal{Y}} |\mathcal{C}(T)|} \qquad (2)$$

For each collection $\mathcal{Y}$ of winning coalitions the formula (2) gives the value of a power index, obtaining in this way the three indices shown below. In the next definitions we assume that $a \in N$ is not a null player because in this case all indices are declared to be zero.

Banzhaf index $(\mathcal{Y} = \mathcal{W}^c)$

$$B_a = \frac{\eta_a(\mathcal{W}^c)}{\sum_{S \in \mathcal{W}^c} |\mathcal{C}(S)|}$$

Holler index $(\mathcal{Y} = \mathcal{W}^m)$

$$H_a = \frac{\eta_a(\mathcal{W}^m)}{\sum_{S \in \mathcal{W}^m} |\mathcal{C}(S)|} = \frac{\eta_a(\mathcal{W}^m)}{\sum_{S \in \mathcal{W}^m} |S|}$$

Alonso–Freixas index $(\mathcal{Y} = \mathcal{W}^s)$

$$AF_a = \frac{\eta_a(\mathcal{W}^s)}{\sum_{S \in \mathcal{W}^s} |\mathcal{C}(S)|} = \frac{\eta_a(\mathcal{W}^s)}{\sum_{S \in \mathcal{W}^s} |S|}$$

4 New Power Indices for Simple Games

In this section we introduce a new model which allows us, following the unified probabilistic scheme shown in the former section, to define three new power indices.

4.1 Model 3

This model is defined by the following assumptions:

- Only winning coalitions in $\mathcal{Y}$ will be formed.
- Coalitions in $\mathcal{Y}$ have a probability of being formed inversely proportional to their size (or cardinality).
- All critical members of the victorious coalition in $\mathcal{Y}$ receive equal shares of the 'spoils', while the rest of players in this coalition receive nothing.

The idea behind this model is the consideration that forming small coalitions is easier and more stable than forming big coalitions.

Items 1 and 3 above are identical as in the previous two models. Thus, the only difference between model 3 and its two predecessors lies on the second item.

Under these assumptions, for each possible election of the set $\mathcal{Y}$, the probability of a coalition $S \in \mathcal{Y}$ to be formed is $P(S) = \dfrac{1/|S|}{\sum_{T \in \mathcal{Y}} 1/|T|}$ and the payoff of a player $a \in S$ is $\dfrac{1}{|\mathcal{C}(S)|} = \chi(S)$ if $a \in \mathcal{C}(S)$, while her payoff is 0 if $a \notin \mathcal{C}(S)$. Thus, the **Expected value of a's gain** is:

$$E_a(\mathcal{Y}) = \sum_{S \in \mathcal{C}_a(\mathcal{Y})} \frac{1/|S|}{\sum_{T \in \mathcal{Y}} 1/|T|} \chi(S) = \frac{1}{\sum_{T \in \mathcal{Y}} 1/|T|} \sum_{S \in \mathcal{C}_a(\mathcal{Y})} \frac{1}{|S||\mathcal{C}(S)|} \quad (3)$$

For each collection $\mathcal{Y}$ of winning coalitions the formula (3) gives the value of a power index, obtaining in this way the three indices shown below. In the next definitions we assume that $a \in N$ is not a null player because in this case all indices are declared to be zero.

Index α ($\mathcal{Y} = \mathcal{W}^c$)

$$\alpha_a = \frac{1}{\sum_{T \in \mathcal{W}^c} 1/|T|} \sum_{S \in \mathcal{C}_a(\mathcal{W}^c)} \frac{1}{|S||\mathcal{C}(S)|}$$

Index β ($\mathcal{Y} = \mathcal{W}^m$)

$$\beta_a = \frac{1}{\sum_{T \in \mathcal{W}^m} 1/|T|} \sum_{S \in \mathcal{C}_a(\mathcal{W}^m)} \frac{1}{|S||\mathcal{C}(S)|} = \frac{1}{\sum_{T \in \mathcal{W}^m} 1/|T|} \sum_{S \in \mathcal{W}_a^m} \frac{1}{|S|^2}$$

Index γ ($\mathcal{Y} = \mathcal{W}^s$)

$$\gamma_a = \frac{1}{\sum_{T \in \mathcal{W}^s} 1/|T|} \sum_{S \in \mathcal{C}_a(\mathcal{W}^s)} \frac{1}{|S||\mathcal{C}(S)|} = \frac{1}{\sum_{T \in \mathcal{W}^s} 1/|T|} \sum_{S \in \mathcal{W}_a^s} \frac{1}{|S|^2}$$

We conclude the section by showing the values of the different power indices considered in this paper in two simple games. The first game is the one defined in Example 1 (Table 1):

Example 1 revisited: Let $(N, \mathcal{W}^m)$ be the simple game, with $N = \{1, 2, 3, 4, 5\}$ and

$$\mathcal{W}^m = \{S \subseteq N : |S| = 3 \quad and \quad S \neq \{3, 4, 5\}\}.$$

Table 1. Power indices for the game in Example 1.

	Model 1	Model 2	Model 3
$\mathcal{W}^c$	$\frac{1}{33}$ $(9, 9, 5, 5, 5)$	$\frac{1}{29}$ $(7, 7, 5, 5, 5)$	$\frac{1}{114}$ $(27, 27, 20, 20, 20)$
$\mathcal{W}^m$	$\frac{1}{27}$ $(6, 6, 5, 5, 5)$	$\frac{1}{27}$ $(6, 6, 5, 5, 5)$	$\frac{1}{27}$ $(6, 6, 5, 5, 5)$
$\mathcal{W}^s$	$\frac{1}{18}$ $(3, 3, 4, 4, 4)$	$\frac{1}{18}$ $(3, 3, 4, 4, 4)$	$\frac{1}{18}$ $(3, 3, 4, 4, 4)$

Example 2. Let $(N, \mathcal{W}^m)$ be the simple game with $N = \{1, 2, 3, 4, 5\}$ and

$$\mathcal{W}^m = \{\{1, 2\}, \{1, 3, 4\}, \{1, 3, 5\}, \{2, 3, 4\}\}.$$

It holds that $5 \succ 4 \succ 3 \succ 2 \succ 1$, $\mathcal{W}^s = \{\{1, 2\}, \{1, 3, 5\}, \{2, 3, 4\}\}$ and

$$\mathcal{W}^c = \{\{12\}, \{123\}, \{124\}, \{125\}, \{134\}, \{135\}, \{234\}, \{1235\}, \{1245\}, \{1345\}, \{2345\}\}.$$

(the voters in each coalition are not separated by commas in this last set) (Table 2).

Table 2. Power indices for game in Example 2.

	Model 1	Model 2	Model 3
$\mathcal{W}^c$	$\frac{1}{63}$ $(25, 19, 11, 6, 2)$	$\frac{1}{25}$ $(9, 7, 5, 3, 1)$	$\frac{1}{192}$ $(70, 56, 36, 22, 8)$
$\mathcal{W}^m$	$\frac{1}{24}$ $(7, 5, 6, 4, 2)$	$\frac{1}{11}$ $(3, 2, 3, 2, 1)$	$\frac{1}{24}$ $(7, 5, 6, 4, 2)$
$\mathcal{W}^s$	$\frac{1}{18}$ $(5, 5, 4, 2, 2)$	$\frac{1}{8}$ $(2, 2, 2, 1, 1)$	$\frac{1}{18}$ $(5, 5, 4, 2, 2)$

5 Ordinal Equivalence of the Three Power Indices Based on Critical Winning Coalitions

It is known [9] that in complete simple games, i.e., games for which any two players are comparable by the desirability relation, the hierarchy in the set of players induced by the Johnston and by the Banzhaf indices coincides with the hierarchy induced by the desirability relation. As it was shown in Sect. 2, both indices are obtained, from model 1 and from model 2 respectively, when $\mathcal{Y} = \mathcal{W}^c$. In this section we prove that the new α index, obtained from model 3 when $\mathcal{Y} = \mathcal{W}^c$, has the same property.

Proposition 5.1. *Let v be a simple game on N and $a, b \in N$. Then,*

(i) $a \succsim b \quad \Rightarrow \quad \alpha_a \geq \alpha_b$.

(ii) $a \succ b \quad \Rightarrow \quad \alpha_a > \alpha_b$.

Proof.

(i) Assume that $a \succsim b$ and let S be a coalition such that b is critical in it ($S \in \mathcal{C}_b(\mathcal{W}^c)$ or $b \in \mathcal{C}(S)$). There are two possibilities:

- If $a \in S$, then $a \in \mathcal{C}(S)$, because $S \setminus \{b\} = (S \setminus \{a, b\}) \cup \{a\} \notin \mathcal{W}$ implies $S \setminus \{a\} = (S \setminus \{a, b\}) \cup \{b\} \notin \mathcal{W}$.
- If $a \notin S$ then a is critical in $T = (S \setminus \{b\}) \cup \{a\}$, because $S = (S \setminus \{b\}) \cup \{b\} \in \mathcal{W}$ implies $T \in \mathcal{W}$, and $T \setminus \{a\} = S \setminus \{b\} \notin \mathcal{W}$. Now, it is $|T| = |S|$ and $|\mathcal{C}(T)| \leq |\mathcal{C}(S)|$. This last inequality is due, on the one hand, to the fact that $a \in \mathcal{C}(T)$, $a \notin \mathcal{C}(S)$, $b \in \mathcal{C}(S)$ and $b \notin \mathcal{C}(T)$, and, on the other hand, because if $x \in \mathcal{C}(T)$ and $x \neq a$ then $x \in \mathcal{C}(S)$. Therefore,

$$\frac{1}{|T||\mathcal{C}(T)|} \geq \frac{1}{|S||\mathcal{C}(S)|}.$$

Thus, it holds

$$\sum_{T \in \mathcal{C}_a(\mathcal{W}^c)} \frac{1}{|T||\mathcal{C}(T)|} \geq \sum_{S \in \mathcal{C}_b(\mathcal{W}^c)} \frac{1}{|S||\mathcal{C}(S)|}, \tag{4}$$

and therefore, $\alpha_a \geq \alpha_b$.

(ii) Assume that $a \succ b$, i.e., $a \succsim b$ and $b \nsucceq a$. As $b \nsucceq a$ there exists $M \subseteq N$ such that $a, b \notin M$, $M \cup \{a\} \in \mathcal{W}$ and $M \cup \{b\} \notin \mathcal{W}$. Thus, a is critical in $T = M \cup \{a\}$, and this coalition T cannot be obtained by either of the two cases analyzed in the first part of the proof. Thus, there is at least one more strictly positive addend in the left part of the inequality (4) than in the right part, and this implies that $\alpha_a > \alpha_b$. $\qquad\square$

Two power indices are said to be *ordinally equivalent in a given simple game* if they induce the same hierarchy on the set of players. Two power indices are said to be *ordinally equivalent in a class of simple games* if they are ordinally equivalent for all games in the class.

The following Corollary is an immediate consequence of Proposition 5.1.

Corollary 5.2. *The Banzhaf index, the Johnston index and the α index are ordinally equivalent in complete simple games.*

6   Conclusion

We have seen that a single probabilistic unified approach serves to generate several families of power indices. Each family has three members according to the kind of winning coalitions considered. The unified approach admits several models that are generated by the probabilities assigned to coalitions. Thus, model 1 contains the Johnston; Deegan and Packel; and Alonso, Freixas and

Molinero power indices, while model 2 contains the Banzhaf, Holler, and Alonso and Freixas power indices.

We have shown that some new models can naturally be defined by illustrating one of them that leads to three new power indices. It is remarkable that the new index based on winning coalitions containing crucial players is ordinally equivalent to the Johnston and to the Banzhaf indices in complete simple games.

Acknowledgements. This research was partially supported by funds from the Spanish Ministry of Economy and Competitiveness (MINECO) and from the European Union (FEDER funds) under grant MTM2015-66818-P(MINECO/FEDER).

References

1. Alonso-Meijide, J.M., Freixas, J.: A new power index based on minimal winning coalitions without any surplus. Decis. Support Syst. **49**(1), 70–76 (2010)
2. Alonso-Meijide, J.M., Freixas, J., Molinero, X.: Computation of several indices by generating functions. Appl. Math. Comput. **219**(8), 3395–3402 (2012)
3. Banzhaf, J.F.: Weighted voting doesn't work: a mathematical analysis. Rutgers Law Rev. **19**(2), 317–343 (1965)
4. Carreras, F., Freixas, J.: Complete simple games. Math. Soc. Sci. **32**(2), 139–155 (1996)
5. Coleman, J.S.: Control of collectivities and the power of a collectivity to act. In: Lieberman, B. (ed.) Social Choice, pp. 269–300. Gordon and Breach, New York (1971)
6. Deegan, J., Packel, E.W.: A new index of power for simple n-person games. Int. J. Game Theory **7**, 113–123 (1978)
7. Einy, E.: The desirability relation of simple games. Math. Soc. Sci. **10**(2), 155–168 (1985)
8. Einy, E., Lehrer, E.: Regular simple games. Int. J. Game Theory **18**, 195–207 (1989)
9. Freixas, J., Marciniak, D., Pons, M.: On the ordinal equivalence of the Johnston, Banzhaf and Shapley power indices. Eur. J. Oper. Res. **216**(2), 367–375 (2012)
10. Holler, M.J.: Forming coalitions and measuring voting power. Polit. Stud. **30**, 262–271 (1982)
11. Isbell, J.R.: A class of simple games. Duke Math. J. **25**(3), 423–439 (1958)
12. Johnston, R.J.: On the measurement of power: some reactions to Laver. Environ. Plann. A **10**(8), 907–914 (1978)
13. Penrose, L.S.: The elementary statistics of majority voting. J. Roy. Stat. Soc. **109**(1), 53–57 (1946)
14. Shapley, L.S., Shubik, M.: A method for evaluating the distribution of power in a committee system. Am. Polit. Sci. Rev. **48**(3), 787–792 (1954)
15. Taylor, A.D.: Mathematics and Politics: Strategy, Voting, Power and Proof, 2nd edn. Springer, New York (2008). https://doi.org/10.1007/978-0-387-77645-3
16. Taylor, A.D., Zwicker, W.S.: Simple Games: Desirability Relations, Trading, and Pseudoweightings. Princeton University Press, New Jersey (1999)

The Story of the *Poor* Public Good Index

Manfred J. Holler[(⊠)]

University of Hamburg, Von-Melle-Park 5, 20146 Hamburg, Germany
Manfred.Holler@uni-hamburg.de

Abstract. The paper starts from the hypothesis that the public good index (PGI) could be much more successful if it were introduced by a more prominent game theorist. It argues that the violation of local monotonicity, inherent to this measure of a priori voting power, can be an asset – especially if the public good interpretation is taken into consideration and the PGI is (re-)assigned to I-power, instead of P-power.

Keywords: Public good index · PGI · Banzhaf index · Shapley-Shubik index · Local monotonicity · I-Power · P-Power · NESS concept

JEL Classification: D71 · C78

1 Why the Public Good Index Is Poor

On the occasion of a workshop on "Institutions, Games and Experiments" at the Max Planck Institute of Economics at Jena, held in honor of Werner Güth's 70th birthday on January 31–February 3, 2014, I had the honor to sit next to Reinhard Selten during two dinners.[1] He told me that he was rather happy to work on a farm in Austria when he was a teenager and his family had to leave Breslau because of the approaching Russian troops. He still liked doing farmwork after his family had to leave Austria because they were German. But as we all know, this was not his final dedication. During the last forty years, Selten and I have met at various conferences and seminars. I remember several intensive discussions during these rare occasions. Yet, I was surprised that he still remembered some of the topics when we were sitting next to each other enjoying our dinners. One topic he remembered was my research on mixed strategy equilibrium. He had warned me that my results might be redundant – and he was right.[2] Another topic was the Public Good Index (PGI). His conclusion was that if a (more) prominent game theorist had introduced this measure of a priori voting power, PGI would be more popular than the Banzhaf index [4] and perhaps as popular as the Shapley-Shubik index [29].

---

[1] Reinhard Selten was born October 5, 1930 in Breslau (today Wrocław), and died August 23, 2016 in Poznań. In 1994, he received a Nobel Prize in Economics together with John Nash and John Harsanyi.

[2] My results were already published in [3], however, this paper was not quoted in the discussion of the mixed strategy equilibrium and its application during 1980–2010.

© Springer-Verlag GmbH Germany, part of Springer Nature 2019
N. T. Nguyen et al. (Eds.): TCCI XXXIV, LNCS 11890, pp. 171–179, 2019.
https://doi.org/10.1007/978-3-662-60555-4_12

Obviously, this was a compliment for the measure, but not necessarily for me. In fact, I have hesitated to talk about Selten's evaluation for more than two years pondering over his resolution. Then I concluded that it should be of general interest to see how much the career of a theoretical concept can depend on the person's status promoting the concept. We should expect that this does not only hold for a theoretical concept, but for scientific work per se. I am sure we can find many examples supporting this hypothesis. For many scientists this seems obvious and needs no further discussion. However, perhaps it does need further discussion because this bias is so obvious – and, in many cases, a serious problem to scientific work. I do not want to give a general analysis of this problem, but outline an example of which I have rather intimate knowledge: the story of the PGI. Section 2 introduces the index, presents some facts of the history of the PGI and outlines the monotonicity problem inherent to this measure. In Sect. 3, the public good interpretation of the PGI will be discussed with respect to the concepts of I-power and P-power. I will argue that the wrong assignment of the PGI, i.e., P-power, leads to confusion and a debasement of this measure. Section 4 concludes this paper with the observation that postulates and stories discriminate between the power measures, however, more generally, the discussion within the community decides about the interpretation and success of an analytical instrument or a theoretical concept.

2   The PGI, Its Prehistory and Its Monotonicity Problem

In their critical review of power measures, Felsenthal and Machover [12: 486] write:

"As far as we know, the first person to be concerned with the measurement of voting power was Luther Martin, a Maryland delegate to the 1787 Constitutional Convention held in Philadelphia. Martin was worried that the voting power of the large states in the US House of Representatives would be disproportionately too large compared to that of the small states, assuming that the representatives of each of the 13 states would always vote as a bloc. In a pamphlet published the following year he not only exposed the fallacy of equating voting power with weight (in this case size of a voting bloc), but made an attempt – albeit unsystematic and somewhat crude – to measure voting power."

Felsenthal and Machover [12: 486] conclude that "Martin's approach is broadly based on the notion of what we have called 'I-power' .... But the measure implicit in his argument apparently relates the voting power of a voter a just to the number of minimal winning coalitions to which a belongs; and therefore seems to us closest to Holler's index...," which is the PGI. In more recent times – almost 200 years after Luther Martin – it seems that I was the first to apply this measure. In [15],[3] I made use of it to evaluate the distribution of voting power in the Finnish Parliament (Eduskunta) underlying the formation of governments for the period 1948–1978. I tried to compare the theoretical power values with the observed participation of the various parties in the governing of

---

[3] This was in the year when Deegan and Packel published their power index referring to minimum winning coalitions, only. See [8]. I am absolutely sure that we did not know of each other, but Riker's "size principle" [28: 32ff] was "in the air." In 1980, I met Ed Packel at his home in Lake Forest, Illinois; I cannot recall that we discussed the "priority issue."

the country.[4] Finland had up to ten parties in its parliament and all parties were in some coalition governments during this period – but the very right party was never in coalition with the very left one, still both were in some government coalitions. During this period, the President of the Republic controlled most of the policymaking[5] – for example, he defined foreign policy and represented the government abroad. He was the commander-in-chief of the armed forces and appointed the members of the government, i.e., the Council of the State. The legislative power was vested with the Parliament. In the exercise of his authority, the President was bound to cooperate with the Parliament through the Council of State, "which must enjoy the confidence of parliament." While there was almost every year a new government, and members of seven different political parties have held position as ministers, Urho Kekkonen was Finnish president from 1956 to 1981.

I had doubts about the PGI because, in this study, this measure produced power values that showed a violation of local monotonicity: party blocs with a larger seat share were assigned smaller power values than party blocs with smaller seat shares. "This property of the MWC index causes doubt concerning its validity" [15: 33]. Note that, instead of PGI, I used the label MWC, i.e., minimum winning coalition – the public good idea did as yet not come into my mind. As an alternative to the PGI, I applied the normalized Banzhaf index which seemed "more adequate in the context of this analysis" [15: 33]. It satisfies local monotonicity.

In order to illustrate the monotonicity problem of the PGI we choose the "notorious example" of a weighted voting game $v = (51; 35, 20, 15, 15, 15)$,[6] i.e., a decision rule $d = 51$ and a vote distribution $w = (35, 20, 15, 15, 15)$ assigned to a set of voters $N = \{1, 2, 3, 4, 5\}$ such that, e.g., $w_2 = 20$. (Voters represent a bloc of votes – the blocs of voters are the players in the voting game v.) The set of minimum winning coalitions of this game (i.e., coalitions that do not contain any surplus players) is

$$M(v) = \{\{1, 2\}, \{1, 3, 4\}, \{1, 3, 5\}, \{1, 4, 5\}, \{2, 3, 4, 5\}\}$$

so that the corresponding PGI is

$$h(v) = \left(\frac{4}{15}, \frac{2}{15}, \frac{3}{15}, \frac{3}{15}, \frac{3}{15}\right).$$

This result indicates that voter 1 is in four minimum winning coalitions of game v, and so on. In its raw version the index counts the number of minimum winning coalitions, i.e., of the "decisive sets" of game v, which have a particular voter i as a member. The numbers 4, 2, 3, 3, and 3 represent the "decisiveness" of the players in game v.

---

[4] See [16] for the empirical results and the analysis.

[5] There were amendments to the constitution and today the power of the President is mostly symbolic. However, officially, the President still leads the nation's foreign politics together with the Council of State and is the Commander-in-Chief of the armed forces.

[6] See [16] for an early discussion of this example.

This counting implies:

(a) Each member of a minimum winning coalition gets the same value, i.e., there is no sharing.
(b) These values are identical for all minimum winning coalitions.
(c) A minimum winning coalition's contribution to the calculation of the PGI is equal to its cardinality.

Independent of the chosen power measure, the set of minimum winning coalitions $M(v)$ fully captures the characteristics of the game v with respect to coalition formation and therefore with respect to power. It is an alternative representation of game v. Therefore, the PGI uses all we know of the game v.

Note that the denominator 15 is the total number of voters who are in some minimum winning coalition of this game. It is chosen for standardizing the PGI so that

$$\sum_i h_i(v) = 1$$

The denominator is not meant to imply a sharing rule. However, this leads us to the interpretation of the index formula. But let us first further discuss the non-monotonicity result illustrated by this simple example above.

Felsenthal and Machover [10: 211] write that "it seems intuitively obvious that if $w_i \leq w_j$ then every voter j has at least as much voting power as voter i, because any contribution that i can make to the passage of a resolution can be equalled or bettered by j." But is j as welcome as i in the forming of coalitions? More generally, do we really *know* that power is monotonic in vote shares? One of the starting points of power index research is the assumption that the vote distribution is a poor proxy for the distribution of voting power. If we could *prove* that the vote-power relation is monotonic, then vote distribution is perhaps a reasonable proxy. However, there is no such proof. Can we trust our intuition?[7]

Felsenthal and Machover [11: 221ff] conclude that "any reasonable power index" should be required to satisfy local monotonicity. Any a priori measure of power that violates local monotonicity is "pathological" and should be disqualified as a valid yardstick for measuring power [11: 221ff]. As a response to this statement, Holler and Napel [20, 21] argue that a violation of local monotonicity could be a characteristic feature of a game, perhaps indicating an instable power relation between the voting blocs with problems to coalition formation. Is the non-monotonicity property of the PGI a pathology or can it serve as an indicator, revealing certain peculiarities of a game?[8] The more popular power measures, i.e., the Shapley-Shubik index and the Banzhaf index, satisfy local monotonicity irrespective of the property of the game.

---

[7] "…if we could trust our intuition, then power indices in general would be rather useless. The number of paradoxes related to the application of these measures, which are the result of a deviation from intuition, indicates that our intuition most likely needs help when it comes to evaluating power – or forming 'reasonable expectations' with respect to power" [18: 607], also see [17].

[8] For further discussion of this argument, see [22, 23].

However, they show non-monotonicities if the vote shares are redistributed. Then there is no guarantee that an increase of a share results in an increase in the (relative) power value. It could well happen that the corresponding power value decreases.[9]

Power is a social phenomenon. It does not only depend on the resources we have but also on the resources of the other decision-makers and the distribution of these resources.

There are many other paradoxes with the Shapley-Shubik index and the Banzhaf index (see, e.g., [7]). Alonso-Meijide and Bowles [1] demonstrate that the Shapley-Shubik index violates local monotonicity if there are a priori unions and, as a consequence, coalitions are no longer equally likely.[10] Note the PGI gives a weight of zero to winning coalitions that contain surplus voters, i.e., voters that are not critical to the winning of a coalition. Thus, coalitions are not equally likely in this case.

Alternatively, we can hope to learn more about weighted voting games if we look for the property of those games which satisfy local monotonicity even when power is measured by the PGI. This is the research program outlined and illustrated in [24]. An interesting variation of this research program has been offered by Freixas and Kurz [14]. They analyze the question how much solidarity we can afford, as measured by the PGI, to guarantee local monotonicity if the "rest of power" is captured by the Banzhaf index. To answer this question, they look for convex combinations of the PGI and the Banzhaf index of a game and their potential of monotonicity. On the one hand, the resulting indices, satisfying local monotonicity, are closer to the Banzhaf than to the PGI, on the other, they are all the more "solidary" than the Banzhaf index.

Earlier, Widgrén [32] proved the following equation that relates the normalized Banzhaf index (β_i) and the PGI (h_i): $\beta_i = (1 - \pi) h_i + \pi \varepsilon_i$. Here π represents the share of winning coalitions that contain surplus players and ε_i represents the share of such coalitions for a particular player i. Widgrén probabilistic model allows to discuss the relationship between the PGI and the Banzhaf index and therefore points to elements which are responsible for the PGI's violation of local monotonicity if observed.

Loosely speaking, the difference between the PGI and the normalized Banzhaf index boils down to those winning coalitions that are not minimal. Holler [16, 19] argues that these coalitions should not be considered because they imply a potential for freeriding if the decisions are on public goods – as is often the case in policy making.[11] This does not mean that surplus coalitions do not form, but they should not be considered when measuring power. The focus on minimum winning coalitions excludes freeriding. If winning coalitions form which contain surplus players, "then it is by luck, similarity of preferences, tradition, etc. – *but not because of power*" [18, 607].

---

[9] This well-known "paradox of redistribution" was introduced in [13].

[10] See [2] for further discussion.

[11] "The *basic principles* underlying the public good index are (a) the public good property, i.e. non-rivalry in consumption and non-excludability of access, and (b) the non-freeriding property. It is obvious from these principles that (strict) minimum winning coalitions should be considered when it comes to measuring power. All other coalitions are either non-winning or contain at least one member that does not contribute to winning. If coalitions of the second type form, then it is by luck or because of similarity of preferences, tradition, etc.—but not because of power, as there is a potential for freeriding" [22: 9].

3 I-Power, P-Power, Solidarity and the Public Good Interpretation

It is not obvious how and when the concept of solidarity emerged to characterize the PGI. Moreover, it is not clear what this concept contributes in addition to the public good interpretation of the PGI introduced in [16].[12] This interpretation was inspired by Brian Barry's [5: 189] observation that the coalition value is a collective good and a concept of dividing the value of a coalition "violates the first principle of political analysis, which is that public policy is a public good (or bad)." For illustration, he continues: "If the death penalty is reintroduced, that pleases those who favour it and displeases those who do not. Similarly, a tax break is a good or bad for people according to their situation. The gains are not confined to those who voted on the winning side nor are the losses confined to those who are on the losing side" [5, 189].

From Paul Samuelson we learned that there is nothing to share in the case of (pure) public goods as both non-rivalry in consumption and non-excludability of access apply. The argument is that those critical to the winning of a coalition exert some power to achieve the public goods which they prefer to alternative outcomes. It is assumed that different minimum winning coalitions bring forward different public goods as collective outcome.

In order to characterize the properties of various indices Felsenthal and Machover (1998) introduced the concepts of I-power and P-power where I-power is meant to capture the influence of (a bloc of) voters on the outcome while P-power summarizes the shares of the spoils voters derive from "winning an election" and holding an office. P-power is about "sharing a cake" while I-power is about having a say in determining the outcome. For instance, [11, 12] suggest that the Banzhaf index describes I-power, an agent's potential influence over the outcome, whereas the Shapley-Shubik index represents P-power, an agent's expected share in a fixed prize. However, as demonstrated by Turnovec [30, 31], with respect to the formal structure of the two measures the distinction does not hold: both measures can be interpreted expressing I-power or, alternatively, P-power. Turnovec convincingly argues that a "power index speaks about the properties of a model, not about the properties of the power as such" [30: 613]. Whether a specific power index is appropriate or not, depends on the properties of the model of collective decision-making which we want to analyse. Therefore, it is not surprising "that there is not just one power index. However, the category in which a specific index falls is not always obvious" [25: 290]. If the model considers the result of the collective decision making regarding a public good, then the sharing approach is inappropriate. Therefore, it is somewhat surprising that Felsenthal and Machover [11] classify the PGI among the P-power measures. It is difficult to see why the PGI does not qualify as an I-power measure like, according to Felsenthal and Machover, the Banzhaf index does. (See the formal relationship of the two measures outlined above.) Moreover, as quoted above, Felsenthal and Machover conclude that Luther Martin's approach is (a) broadly based on the notion of I-power, and (b) it seems closest to

---

[12] In the sequel, Holler and Packel [26] axiomatized the PGI. Napel [27] completed this axiomatization.

"Holler's index" - which is the PGI. It is therefore inconsistent assigning the PGI to P-power.

This assignment is of course not without consequences. In the case of P-power and sharing a cake, the violation of local monotonicity could be a problem and perhaps earn the label pathology. But does this still hold in case that we want to measure the influence of blocs of voters on the outcome? There is a lot of empirical evidence that this influence might be non-monotonic – like in the case of the Finnish Parliament (see above).[13]

It seems natural that the PGI does not achieve good results if tested with respect to postulates stated for P-power. Felsenthal [9: 368] gives a substantial list of six postulates that a "reasonable P-power index should satisfy." We read

"...the various indices proposed to date for estimating the expected share in the fixed prize of the members in an *n*-person cooperative game – which, following Felsenthal and Machover (1998, ch. 6), we shall call *P-power indices* – must be assessed by examining which postulates – i.e., intuitively compelling conditions – they satisfy. Failure to satisfy these conditions is a suspect counter-intuitive behavior, which can be regarded as paradoxical or, in extreme cases, pathological, and may indicate that a P-power index guilty of it must be discarded" [9: 368].

In the list of P-power indices analyzed by Felsenthal we find the PGI which violates four of the six postulates that Felsenthal proposed for P-power indices – and therefore should be discarded. Indeed, the PGI should be discarded from the list of P-power indices as it is not meant to be a P-power index. One could have argued that this poor result should have signaled that the PGI does not represent P-power. The underlying public good model does not support sharing. It seems that Felsenthal gets rather close to this insight:

"But the prize of victory according to the PGI index is regarded not as a unit of TU, a private good, to be divided among the members of a victorious MWC, but as a public good enjoyed in its entirety by all members of this coalition (but only by them!)" [9: 377f].

One should add that some people, not in the coalition, will be lucky and also enjoy this public good X, but they have no influence on whether X is produced or not. For criminals this luck has, however, a "negative sign" in the case that additional police officers are hired for increasing the security of the citizens.

Public goods are a close relative to externalities and therefore invite freeriding. Consequently, if we relate power to causality, we should exclusively consider decision-makers which are decisive for their production, i.e., "necessary elements of a sufficient set" – which justifies to focus on minimum winning coalitions only. The NESS-concept ("necessary elements of a sufficient set") has been introduced in philosophy to discuss causality in collective actions.[14] The set of the "necessary elements of a sufficient set" is a minimum winning coalition. This suggests the PGI as measure of causality. However, irritated by non-monotonicity results, Braham and van Hees propose a *weak*

---

[13] In this case, one might argue that the situation allows for an interpretation that considers P-power. Under the umbrella of the power of the President, the numerous coalition governments were sharing the benefits of holding an office. The contribution to the political result was perhaps secondary to many participants in this game.

[14] See [6] with reference to [33].

NESS-test ending up with the Banzhaf index instead of the PGI (which corresponds to the *strong* NESS-test).

4 Concluding Remarks

The discrimination against non-monotonicity, implicit in the weak NESS-test, ignores that possible causal relations are not only determined by the resources of an agent but also depend on the resources of the other agents and how they are employed. In the case of power, the weak NESS-test does not consider that power is a social phenomenon and therefore, in general, not only depending on the resources of a single decision maker.

We have seen that the objections to the PGI because of its non-monotonicity results are closely related to its P-power assignment. There is nothing in the mathematical structure of the various indices which says whether the index expresses sharing influence (I-power) or sharing a cake (P-power). There are the underlying stories and the scientific community that put the measures in the various characterizing boxes. Still, a concept can get misplaced – and doomed. It is the discussion within the community that decides on its performance and its status. I hope that this paper helps that the PGI is no longer discarded for producing pathologies, but applied clarifying the notion of power in cases of collective decision-making when the outcome can be considered a public good.

References

1. Alonso-Meijide, J.M., Bowles, C.: Power indices restricted by a priori unions can be easily computed and are useful: a generating function-based application to the IMF. Ann. Oper. Res. **137**, 21–44 (2005)
2. Alonso-Meijide, J.M., Bowles, C., Holler, M.J., Napel, S.: Monotonicity of power in games with a priori unions. Theor. Decis. **66**, 17–37 (2009)
3. Aumann, R.J., Maschler, M.: Some thoughts on the minimax principle. Manag. Sci. **18**, 54–63 (1972)
4. Banzhaf III, J.F.: Weighted voting doesn't work: a mathematical analysis. Rutgers Law Rev. **19**, 317–343 (1965)
5. Barry, B.: Is it better to be powerful or lucky? Parts 1 and 2. Polit. Stud. **28**, 183–194 and 338–352 (1980)
6. Braham, M., van Hees, M.: Degrees of causation. Erkenntnis **71**, 323–344 (2009)
7. Brams, S.J.: Game Theory and Politics. Free Press, New York (1975)
8. Deegan, J., Packel, E.W.: A new index of power for simple n-person-games. Int. J. Game Theory **7**, 113–123 (1978)
9. Felsenthal, D.S.: A well-behaved index of a priori p-power for simple n-person games. Homo Oeconomicus **33**, 367–381 (2016)
10. Felsenthal, D.S., Machover, M.: Postulates and paradoxes of relative voting power - a critical appraisal. Theor. Decis. **38**, 195–229 (1995)
11. Felsenthal, D.S., Machover, M.: The Measurement of Voting Power. Theory and Practice, Problems and Paradoxes. Edward Elgar, Cheltenham (1998)

12. Felsenthal, D.S., Machover, M.: Voting power measurement: a story of misreinvention. Soc. Choice Welfare **25**, 485–506 (2005)
13. Fischer, D., Schotter, A.: The inevitability of the paradox of redistribution in the allocation of voting weights. Public Choice **33**, 46–67 (1980)
14. Freixas, J., Kurz, S.: The cost of getting local monotonicity. Eur. J. Oper. Res. **251**, 600–612 (2016)
15. Holler, M.J.: A priori party power and government formation. Munich Soc. Sci. Rev. **1**, 25–41 (1978). (republished in: Holler, M.J. (ed.) Power, Voting, and Voting Power. Physica-Verlag, Würzburg and Wien (1982); and Munich Social Science Review, New Series, vol. 2 (2018))
16. Holler, M.J.: Forming coalitions and measuring voting power. Polit. Stud. **30**, 262–271 (1982)
17. Holler, M.J.: The public good index. In: Holler, M.J. (ed.) Coalitions and Collective Action. Physica-Verlag, Würzburg and Vienna (1984)
18. Holler, M.J.: Power, monotonicity and expectations. Control Cybern. **26**, 605–607 (1997)
19. Holler, M.J.: Two stories, one power index. J. Theor. Polit. **10**, 179–190 (1998)
20. Holler, M.J., Napel, S.: Local monotonicity of power: axiom or just a property. Qual. Quant. **38**, 637–647 (2004)
21. Holler, M.J., Napel, S.: Monotonicity of power and power measures. Theor. Decis. **56**, 93–111 (2004)
22. Holler, M.J., Nurmi, H.: Reflections on power, voting, and voting power. In: Holler, M.J., Nurmi, H. (eds.) Power, Voting, and Voting Power: 30 Years After, pp. 1–24. Springer, Heidelberg (2013). https://doi.org/10.1007/978-3-642-35929-3_1
23. Holler, M.J., Nurmi, H.: Pathology or revelation? The public good index. In: Fara, R., Leech, D., Salles, M. (eds.) Voting Power and Procedures. SCW, pp. 247–257. Springer, Cham (2014). https://doi.org/10.1007/978-3-319-05158-1_14
24. Holler, M.J., Ono, R., Steffen, F.: Constrained monotonicity and the measurement of power. Theor. Decis. **50**, 385–397 (2001)
25. Holler, M.J., Owen, G.: On the present and future of power measures. Homo Oeconomicus **19**, 281–295 (2002); republished as Present and future of power measures. In: Kacprzyk, J., Wagner, D. (eds.) Group Decisions and Voting, pp. 31–46. Academicka Oficyna Wydawnicza, Warsaw (2003)
26. Holler, M.J., Packel, E.W.: Power, luck and the right index. Zeitschrift für Nationalökonomie (J. Econ.) **43**, 21–29 (1983)
27. Napel, S.: The holler-packel axiomatization of the public good index completed. Homo Oeconomicus **15**, 513–521 (1999)
28. Riker, W.H.: The Theory of Political Coalitions. Yale University Press, New Haven (1962)
29. Shapley, L.S., Shubik, M.: A method of evaluating the distribution of power in a committee system. Am. Polit. Sci. Rev. **48**, 787–792 (1954)
30. Turnovec, F.: Power, power indices and intuition. Control Cybern. **26**, 613–615 (1997)
31. Turnovec, F.: Power indices: swings or pivots? In: Wiberg, M. (ed.) Reasoned choices. Essays in Honour of Academy Professor Hannu Nurmi on the Occasion of his 60th birthday. Digipaino, Turku (2004)
32. Widgrén, M.: On the probabilistic relationship between the Public Good Index and the Normalized Banzhaf index. In: Holler, M.J., Owen, G. (eds.) Power Measures, vol. 2. Homo Oeconomicus **19**, 373–386 (2002)
33. Wright, R.: Causation, responsibility, risk, probability, naked statistics, and proof: Pruning the bramble bush by clarifying the concepts. Iowa Law Rev. **73**, 1001–1077 (1988)

Author Index

Algaba, Encarnación 90

Bertini, Cesarino 109
Bezzi, Mirko 146
Buczek, Aleksander 74

Cofta, Piotr 127

Dall'Aglio, Marco 35

Flis, Jarosław 1
Fragnelli, Vito 35, 90
Freixas, Josep 162

Gambarelli, Gianfranco 109, 146

Holler, Manfred J. 21, 171

Llorca, Natividad 90

Mercik, Jacek 74
Moretti, Stefano 35
Motylska-Kuźma, Anna 74

Nurmi, Hannu 63

Orłowski, Cezary 127

Pastuszka, Józef 127
Platje, Johannes (Joost) 47
Pons, Montserrat 162

Rupp, Florian 21

Sánchez-Soriano, Joaquin 90
Słomczyński, Wojciech 1
Stach, Izabella 109
Stolicki, Dariusz 1

Wąsik, Mariusz 127
Welfler, Piotr 127

Zibetti, Giuliana 109
Zibetti, Giuliana Angela 146

Printed in the United States
By Bookmasters